Synergized

Middle School Chemistry

Matter's Phases and Properties

Sharon F. Johnson Ph.D.

Joanne J. Smith M.A.

ISBN: 1456329707
ISBN-13: 9781456329709

Preface

The challenges for middle school science teachers are greater than ever before. Teachers frequently struggle with issues such as:

1. How do I find the time to design meaningful, in-depth learning experiences?
2. How do I find resources to effectively teach middle level physical science, especially chemistry?
3. How do I integrate literacy into science lessons?
4. How do I assess student learning?
5. How do I increase student achievement?
6. How do I engage and excite kids about science?
7. How do I incorporate science instructional best practices?
8. How do I differentiate for special needs students?

Synergized Middle School Chemistry (SMSC) provides the middle school science educator (grades 5-9) with the curriculum to answer these questions and ease the demands of teaching. *SMSC's* tried and tested lessons are economical, comprehensive, and exciting. *SMSC* students actively participate in engaging, hands-on investigations. As students read, write, calculate, and communicate in the context of science, they not only deepen their conceptual understanding but also improve their literacy and math skills. Feedback for students is continuous and ongoing. To save both time and money, synergized investigations use safe, economical supplies found in supermarkets and discount stores. This makes *SMSC* very adaptable to both classroom and home-based learning settings.

The *SMSC* CURRICULUM

SMSC: Matter's Phases and Properties (Volume 1) is a teacher's guide. It includes all *Student Lessons* and *Teacher Notes* related to the properties of solids, liquids, and gases, kinetic molecular theory, density, and intermolecular attractive forces.

SMSC: Elements and Interactions (Volume 2) is a teacher's guide. It includes all *Student Lessons* and *Teacher Notes* related to the structure of atoms & molecules, elements, the periodic table, physical and chemical changes, and acids and bases.

SMSC: Student Edition: Matter's Phases and Properties & Elements and Interactions contains all the *Student Lessons* from Volumes 1 & 2 as well as an introduction and appendix written especially for students. The student edition does not contain *Teacher Notes*. The *Teacher Notes* in *SMSC Volumes 1 & 2* include lesson rationales, teaching strategies, materials preparation, background and sample student responses, and post-assessments that differ from the practice assessments in the *SMSC Student Edition*.

Dedication

This book is dedicated to the many middle level science teachers who encouraged us to pursue this endeavor. We were inspired by their commitment to provide meaningful science experiences for their students. Our hope is that this book fulfills their needs and eases the demands of today's teachers.

Acknowledgments

We wish to thank colleagues, friends, and family for their wisdom, insights, advice, and support. We are synergetically yours.

Doris Kimbrough, Associate Professor of Chemistry, College of Liberal Arts and Sciences, University of Colorado at Denver, for her chemical genius, incredible ability to explain concepts in "middle school words," entertaining demonstrations, and fantastic salty stories.

Doug Smith, Chemistry Instructor, Colorado Mountain College, Steamboat Springs, for his insightful editorial advice, never ending confidence in us, and tolerance for life east of the tunnel.

Bill Johnson, for his unconditional belief in our work, infinite patience, and photographic good looks.

Elnore Grow, Chemistry Educator, for her unselfish devotion to Colorado chemistry education, wizardly ways, and continuing friendship.

Linda Block-Gandy, Science Education Consultant, for sharing her L.I.P.S strategies, the diamond rule, and unmatched sense of humor.

Linda Fiorella, Literacy Consultant, for her literacy expertise, encouragement, and work breaks in the 'boat.

Scott Wallace, Teacher, Englewood Middle School, for his down-to-earth teaching methods, out-of-this world expertise, and explosive personality.

The BSI Team: Julie, Lisa, Shaun, Tim, and Cyndie for inspiration and the opportunity to try out our ideas.

Carole Basile, Associate Professor, School of Education and Human Development, University of Colorado at Denver, for her guidance, leadership, and note-worthy insights.

Bill Timpson, Professor, School of Education, Colorado State University, for his practical advice, teaching tips, and 'yes, you can attitude' when we weren't sure we could write a book.

Coal Creek Elementary School Public Art Project, Louisville, Colorado, for the cover photo, of their students' hands.

About the Authors

Sharon F. Johnson received a Bachelor of Science in the Teaching of the Biological and General Sciences, a Master of Science in the Teaching of Science, and a Ph.D. in Education from the University of Illinois, Urbana – Champaign. She is currently working as an educational consultant and author. For the past six years, she has been a pedagogy instructor for the chemistry courses taught through the NSF sponsored Rocky Mountain Middle School Math and Science Partnership at the University of Colorado at Denver. From 1968 to 1994, Sharon taught middle school and high school science in Illinois, Florida, and Colorado. From 1994 to 2004, she was the science curriculum specialist for the Adams Twelve Five Star School District in Colorado.

Sharon F. Johnson was the lead facilitator for the Adams Twelve Schools and the University of Colorado at Denver CONNECT, a National Science Foundation Systemic Initiative for Mathematics and Science. She co-authored and coordinated The Using Literacy Integration to Communicate Scientifically Project (ULINCS), a state funded partnership between the University of Northern Colorado and Adams Twelve schools. She has presented at both national and state science conventions on the topics of inquiry-based science, assessment, and the science-literacy connection.

Her writings include a set of activity guides for the University of Colorado at Boulder Science Discovery Program highlighting their CU Wizards presentations. Her latest work is a chapter co-authored for the Association of Teacher Educators, ATE Yearbook XVIII entitled, "Capturing Teacher Learning, Curiosity, and Creativity through Science Notebooks."

Joanne J. Smith received her Bachelor of Science in Botany from Colorado State University and her Master's of Arts in Special Education: Gifted and Talented from the University of Northern Colorado. She is currently working as an educational consultant and author. For the past five years, she has been an instructor for the Biological Sciences Initiative – Ready, Set, Science Program at the University of Colorado at Boulder providing workshops and mentoring opportunities for beginning middle school science teachers. From 1976 to 2004, Jo taught middle school and high school science in Colorado.

Joanne has been honored throughout her teaching years for her work in science education. She received the Colorado Excellence in Teaching Science Award in 1989, awarded an Access Excellence Fellowship in 1996, was a member of the Pfizer Leadership Institute in Molecular Genetics at Cold Spring Harbor, and received the Radio Shack National Teacher Award in 2002.

Joanne has presented at both national and state science conventions on the topics of assessment, calculator-based lab training, and using portfolios in the science classroom. She served on the Colorado Association of Science Teachers Board as a secondary regional director. Jo has been a writer and contributor to a number of curriculum projects related to assessment, literacy, and inquiry-based instruction.

Table of Contents

Introduction

Synergized is a word you won't find in the dictionary. It is a word created by melding the words "synergy" and "energy." Synergized lessons work together in a prescribed instructional sequence to teach core concepts more effectively than a series of unconnected lessons. The synergy comes from integrated, meaningful interactions between teachers and students.

Synergized Middle School Chemistry (SMSC) offers an exciting new approach for teaching middle school physical science concepts. Synergetic lessons incorporate the best practices of science instruction: inquiry-based learning, literacy integration, and embedded assessments. The book includes practical advice for teachers about content, material preparation, and instructional strategies.

THE CONTENT: MATTER'S PHASES AND PROPERTIES

This book is divided into three sequential lesson sets. The lesson sets are designed around the national content standards for middle school physical science. The concepts are selected and organized to confront and dispel common chemistry misconceptions as well as build a strong chemistry background for further study in high school.

Lesson Set One: Matter's Phases

Students investigate solids, liquids, and gases. They study phase changes associated with melting, freezing, evaporating, and condensing. Kinetic theory is introduced and used to describe the behavior of atoms and molecules in each phase of matter. Throughout the lesson set, students are prompted to relate their macroscopic observations of matter to the submicroscopic behavior of atoms and molecules. Note: A safety lesson is also included prior to students using chemical materials and equipment.

Lesson Set Two: Density

Students define density and identify it as a characteristic property of matter. They calculate density values from laboratory measurements. Students use water displacement to calculate the volume of irregular solids. Students identify factors that influence density values. They observe and compare the densities of several common substances. Students apply the concept of density to a real world issue.

Lesson Set Three: Attractive Forces

Students learn about the attractive forces that hold atoms and molecules together to form solids and liquids. They study water's behavior in a variety of situations and learn how both adhesive and cohesive forces play a role in explaining surface tension. The effect of surfactants is explored and investigated.

Correlation to the **National Science Education Standards** (National Research Council, 1996, pp. 104-111)	Phases of Matter	Density	Attractive Forces
Unifying Concepts and Processes			
Systems, order, and organization	X	X	X
Evidence, models, and explanation	X	X	X
Constancy, change and measurement	X	X	X
Science as Inquiry Content Standard A – Levels 5-8			
Abilities Necessary to do Scientific Inquiry	X	X	X
Understandings About Scientific Inquiry	X	X	X
Physical Science Content Standard B – Levels 5-8			
Properties and Changes of Properties in Matter	X	X	X
Motions and Forces	X	X	X
Transfer of Energy	X	X	X
Physical Science Content Standard B – Levels 9-12			
Structure of atoms			
Structure and Properties of Matter	X	X	X
Chemical reactions			
Motions and Forces			
Conservation of energy and increase in disorder	X	X	X
Interactions of energy and matter	X	X	X
Science and Technology Content Standard E – Levels 5-8			
Abilities of Technology Design	X	X	X
Understanding about science and technology	X	X	X
Science in Personal and Social Perspectives Content Standard F – Levels 5-8			
Personal Health	X	X	
Populations, Resources, and Environments		X	
Risks and Benefits	X	X	
Science and Technology in Society	X	X	X
History and Nature of Science Content Standard G – Levels 5-8			
Science as a human endeavor	X	X	X
Nature of science	X	X	X
History of science		X	

SYNERGIZED INQUIRY-BASED SCIENCE INSTRUCTION

Much has been written about inquiry-based instruction. Our work is based on the National Research Council's (NRC) book, *Inquiry and The National Science Education Standards*. The NRC states, "The content standards for Science as Inquiry include both abilities and understandings of inquiry (NRC, 2000, p. 18)." The Council goes on to list the fundamental inquiry *abilities* targeted for grades five through eight.

- Identify questions that can be answered through scientific investigations.
- Design and conduct a scientific investigation.
- Use appropriate tools and techniques to gather, analyze, and interpret data.
- Develop descriptions, explanations, predictions, and models using evidence.
- Think critically and logically to make the relationships between evidence and explanations.
- Recognize and analyze alternative explanations and predictions.
- Communicate scientific procedures and explanations.
- Use mathematics in all aspects of scientific inquiry. (NRC, 2000, p. 19)

In order to master the fundamental inquiry abilities, students need extensive opportunities to learn and practice the skills. Thus, each *SMSC* investigation targets specific inquiry abilities in a number of ways. Focus questions frame investigations. Predictions and/or hypotheses are made. Observations and data are recorded and analyzed to make claims based on evidence. At the end of the investigations, questions prompt students to relate new understandings and reconcile possible misconceptions with new content. And, every investigation emphasizes both oral and written communication skills.

SMSC investigations foster critical thinking. Students are asked to state claims based on evidence and then share their claims with each other and the class. When claims conflict, students are prompted to critically analyze the evidence and rethink their claims.

Most of the investigations are guided inquiries. Some are open inquires. Guided inquiries provide opportunities for students to learn and practice their inquiry abilities; open inquiries give students the chance to apply what they have learned to new situations. Open inquiries also provide students with the experience of designing and conducting their own investigations. Both guided and open inquiries are used to monitor and assess student inquiry abilities.

Investigations require students to apply mathematics as they take measurements, organize, and analyze data. Number sense is stressed through data analysis and the evaluation of evidence to support claims.

The second part of the inquiry standards, as defined by the NRC, lists the following *understandings* for grades five through eight.

- Different kinds of questions suggest different kinds of scientific investigations.
- Current scientific knowledge and understanding guide scientific investigations.
- Mathematics is important in all aspects of scientific inquiry.
- Technology used to gather data enhances accuracy and allows scientists to analyze and quantify results of investigations.

- Scientific explanations emphasize evidence, have logically consistent arguments, and use scientific principles, models, and theories.
- Science advances through legitimate skepticism.
- Scientific investigations sometimes result in new ideas and phenomena for study, generate new methods or procedures for an investigation, or develop new technologies to improve the collection of data. (NRC, 2000, p. 20)

Teachers will find many opportunities to address these understandings as students perform *SMSC* lessons. For example, often times the measuring tools available do not provide consistent data from group to group. This is an excellent time to discuss the shortcomings of measurement tools and how both accuracy and precision affect scientific work.

Student inquiries play a major role in understanding how one investigation leads to another. Most of the time student inquiries generate more questions than answers. Students need to know that scientific inquiry works the same way for them as it does for scientists. Whenever possible, modern day examples of how the inquiry process produces new knowledge and additional questions should be referenced and discussed.

Synergetic science motivates students to achieve at higher levels by structuring learning around inquiry-based abilities and understandings. Inquiry, whether guided or open-ended, allows students to construct conceptual understanding through direct experience and thoughtful instruction. The use of inquiry in *SMSC* lessons centers on the belief that students who are active and engaged learners will achieve at higher levels than those who are passive and uninvolved in their learning. But, for instruction to be synergized, inquiry must be integrated with content objectives, literacy and assessment.

SYNERGIZED LITERACY INSTRUCTION

Teaching is a strategic act. The pedagogical strategies chosen by teachers have a significant impact on student learning (Marzano, 2001). Strategies are not taught or learned in one class. They take time to become internalized by the teacher and make sense to the students. And, as in all teaching, implementing a strategy well is an artful endeavor.

Our lessons include literacy strategies to teach scientific vocabulary and nonfiction reading. Scientific writing is developed through the use of science notebooks.

Teaching Reading in the Science Classroom

The ultimate goal of reading instruction is to produce students who are strategic readers. If taught well, students should be able to read on their own with comprehension and fluency using the strategies learned in their classes.

"Who should teach reading?" The answer is simple, "Everyone!" Reading is not the sole responsibility of the reading or language arts teacher. For reading strategies to be effective, they must be used in cross-curricular settings and outside the classroom. Each content area has its unique vocabulary and text structure. Thus, each science teacher has the responsibility for teaching students the language of science, how to navigate nonfiction text, read for information, and think critically about the content. Without these skills, students will not be able to achieve at higher levels.

Many assessment experts note a positive correlation between student reading scores and science scores on large-scale assessments. For students to show gains on standardized assessments, teaching students to read and comprehend nonfiction text is imperative. To this end, students need to have access to a variety of nonfiction readings found in magazines, newspapers, trade books, and standard references as well as up-to-date textbooks. Readings should be selected based on their appropriateness for adolescents. School reading resource teachers and media specialists can help with selecting appropriate materials for varying reading levels. Ideally, every science teacher should have a classroom library to meet the needs of all learners.

Unfortunately, most science teachers receive very little instruction on how to effectively integrate literacy into their science lessons. Thus, students become frustrated by the dense text and difficult vocabulary. Assignments go unread or not understood. In response, science teachers give fewer assignments and spend less time using text information to build conceptual understanding. In turn, fewer students gain the scientific literacy necessary to work in a scientific field or understand scientific advancements.

Synergetic Reading Strategies

There are many excellent books available on the teaching of nonfiction reading. Most describe a variety of strategies to improve scientific vocabulary and reading comprehension. Having classroom tested a variety of reading strategies, we have found that some work better than others for middle school science. Strategies that work the best have the following features:

- Build on the student's previous hands-on experience with the science concepts
- Allow for individual, small group, and whole class discussion
- Allow every student to have a successful reading experience
- Build reading confidence
- Provide immediate, positive feedback
- Have an element of fun or novelty
- Involve collaboration with others
- Emphasize active, participatory reading much like a hands-on lab
- Provide speaking and listening practice
- Involve summarizing ideas
- Provide opportunities to write about what was read
- Emphasize scientific vocabulary building
- Increase understanding of core concepts
- Apply new vocabulary and core ideas

We have also found that the best strategies are those that students self-select to use for independent reading. For example, if a student is taught that good science readers often read a single paragraph, stop, and then reread the paragraph to determine the main idea, they will have a valuable skill they can use on their own. Thus, the job of reading instruction is not just to learn from the immediate reading but also to provide the practice needed to use the skill independently in the future. To this end, we've found that using a few strategies over and over is better than using many strategies once or twice.

Every *SMSC* reading is paired with at least one selected reading strategy. And, each reading is accompanied with teacher notes that detail how to use the strategy. In the literacy resource section at the end of the book, each strategy is described along with reference sources.

Using Science Notebooks to Improve Student Writing Skills

A student science notebook is the perfect writing tool. Just as a working scientist's notebook is an essential tool for scientific research, so the student science notebook is an essential tool for inquiry into science concepts. Notebooks give students a reason to write; notebooks authenticate the student writing process. Our lessons are designed for use with a student science notebook. It is assumed that students will record their work in their notebook. Science notebooks allow students to organize science information for daily and future use. In notebooks students write their questions, predictions, observations, record and display data, state claims and evidence as well as their new understandings. Thus, notebooks provide practice for literacy-based skills like note taking and communicating their ideas through writing.

Learning how to use a notebook is a basic science inquiry skill. Student science notebooks are most effective when used as part of the daily classroom routine. Once students become familiar with the notebook format, they can focus on lesson content and learning is enhanced.

SYNERGIZED ASSESSMENT

Different assessment formats provide different kinds of information used for different purposes. Classroom and laboratory assessments focus on learning by providing feedback to students, while international and national assessments provide data for system accountability (Doran, 2002, p 10).

Assessment is an essential part of the educational process. Synergetic science assessments evaluate both the student's conceptual understanding of chemistry and the student's ability to apply and understand inquiry. Research has shown that regular, high quality assessment in the classroom can have a positive effect on student achievement. Synergized assessments have the following attributes:

- embedded and integrated throughout the curriculum,
- set in a variety of contexts,
- developmentally appropriate,
- can be modified to accommodate students with diverse needs, and
- provide opportunities for students to evaluate and reflect their scientific understandings and abilities.

SYNERGIZED LESSONS

SMSC consists of sequential lessons designed to maximize student engagement and learning. Each lesson set contains the following types of lessons.

Student Investigations

> Different kinds of questions suggest different kinds of scientific investigations. Some investigations involve observing and describing objects or events; ...some involve experiments; some involve seeking more information; some involve the discovery of new objects and phenomena; and some involve making models (NRC, 1996, p. 148).

As stated in the inquiry section, *SMSC* investigations are aligned to the abilities and understandings of inquiry. Students are provided a focus question, asked to predict, and follow a procedure. Next, they observe, collect, and display data; they use their observations and data to state claims and evidence. Finally, they write about their new understandings and reflect on their work.

The *claims and evidence* section along with the sections called *new understandings* and *reflections* replace the traditional conclusion found in many lab report formats. In *new understandings*, students tell what else they have learned, apply their learning to new situations, and evaluate the design of the investigation. The *reflections* section allows students to think and write about their learning experiences as a whole. Teachers may add a conclusion to the investigation, if needed.

SMSC investigations have been thoughtfully designed to use safe, obtainable, and inexpensive materials and supplies. With only a few exceptions, materials can be purchased at grocery, pharmacy, or other local stores. A safety lesson is found in Matter's Phases: Lesson Two, prior to students using chemical materials and equipment.

Chart Reading, Graphing and Data Interpretation

Students collect and analyze data, construct graphs, and draw conclusions. They use scientific data presented in charts and graphs to reinforce concepts presented in the lessons. Since many standardized tests use graph and chart reading to assess student learning, the lessons provide additional practice to develop these skills.

Teacher Demonstrations

Teacher demonstrations are designed to introduce, excite, and extend student learning. Detailed teacher notes accompany each demonstration.

Concept Readings

Readings explain science concepts. They are an integral part of the lesson sequence and provide critical information. Literacy strategies are specifically selected for each reading to enable students to learn how to read nonfiction with comprehension and fluency. Many of the literacy strategies focus on developing student vocabulary.

The chemistry readings are written with the assumption that students do not have access to textbooks or other outside readings; however, if other readings are available, they most certainly should be considered for use. Synergetic readings can be reproduced for classroom use. Few pictures are used to keep publishing costs down. Of course, illustrations, photographs, and drawings are important text features that enrich understanding. If available, the teacher should use other print materials to supplement the basic concept readings in each lesson set.

Assessments

Synergetic lessons use a variety of classroom assessment formats to evaluate student learning. In the lesson set overviews, a teacher assessment guide identifies classroom assessments for use before, during, and at the end of the lessons. Specific activities and checkpoints are provided to help teachers uncover student ideas and check for new understandings. At the end of each lesson set, a post-assessment provides information for both student evaluation as well as system accountability.

Pre-Assessments

Each lesson set begins with a formal pre-assessment designed to uncover student ideas. These lessons also introduce the investigations that follow. During the pre-assessment, the teacher's role is to ask clarifying questions to identify student pre-conceptions and background knowledge. At this time, teachers should not explain the science. As student questions emerge, their questions should be recorded, posted, and/or referenced.

Although each lesson set begins with a formal pre-assessment, all lessons informally pre-assess student ideas. These pre-assessment opportunities take the form of focus questions, predictions/hypothesis, class discussions, literacy strategies, and reflections. The teacher's role is to use the ideas generated by the students to guide further instruction and modify lessons accordingly.

Science Notebooks as an Assessment Tool

Science notebooks provide a valuable insight into student learning. For example, at the end of investigations, students are prompted to write about their new understandings. This is a critical part of concept development. It also provides valuable informal assessment information for the teacher. *SMSC* prompts are designed to solicit key information related to student learning as well as provide an opportunity for open-ended responses.

The claims and evidence sections of each investigation assesses whether or not students are able to link their observations and data to the focus question and make valid claims. Reading selected claims and evidence responses, as indicated in the assessment matrix, is an effective method to evaluate student learning.

The assessment matrix highlights opportunities to use science notebook entries for student evaluation. Entries that will be read and/or evaluated by the teacher need to be clearly communicated to the student before each lesson.

Review Lessons

Review lessons bring together major concepts and experiences from previous lessons. They extend learning by reinforcing concepts in new and different ways. These lessons provide valuable diagnostic information prior to administering the post-assessment. If it becomes evident that students need more instruction, teachers should take the time to re-teach and/or revisit key concepts.

Post-Assessments

Each set of lessons includes a post-assessment to evaluate student learning. *SMSC* assessments use selected and constructed response items. Scoring guides are provided. In addition, performance-assessments opportunities are noted in the teacher notes.

TEACHER SUPPPORT
Lesson Set Overviews

The overview, found at the beginning of each unit, includes the lesson concepts, key vocabulary, lab equipment and supplies, a synopsis of each lesson, and an assessment guide identifying pre-assessment, on-going, and post-assessment opportunities.

Detailed Teacher Notes

Detailed teacher guidance is provided for each lesson. The lesson rationale, materials preparation instructions, teacher's role, teacher's background information, and teaching tips are included. Sample student responses and/or scoring guides are provided as appropriate. Teacher notes give helpful hints and suggestions for Internet sites, optional extension activities, and related resources.

Managing the Synergized Classroom: Tips and Techniques

SMSC lessons require students to take responsibility for their own learning. For this to happen, the teacher must maintain a productive, safe learning environment with clear expectations and follow through. The instructor's responsibility is to provide a student-centered environment where respect is paramount. The following are tips and techniques for managing a successful synergized science classroom.

Establish a purpose for learning. Let students know they will be actively involved in the process of science. Students should be aware that the curriculum in this course sets the foundation for future science classes and that the content they will be studying is applicable to their daily lives.

Validate students, their learning, and their time. Synergetic lessons offer many opportunities to validate student work both formally and informally. Students thrive on positive reinforcement.

Be prepared and organized. Plan at least a week at a time and adjust daily for slight changes. Pack the class period with relevant instructional activities. Minimize student free time as it can trigger behavioral issues.

Set clear expectations. Students need to know what materials are expected daily for class, how and when to make-up missed work, how to clean up after a lab, and how to perform science labs safely.

Be consistent. Send the same message to all students. Treat individual issues privately.

Begin class on time. This is when you have the students' attention – they need to be productive the minute class starts. Save attendance and housekeeping chores for a time when students are engaged in their learning. Set a routine and follow it daily.

Use a seating chart. Seating charts designed with a purpose are critical for a productive synergized classroom. Make changes in the seating chart to allow different groups of students to work together, to separate nonproductive classmates, and to allow for other issues that may arise (physical needs such as seeing the board or hearing problems).

Purposely assign lab teams. Group students to accomplish particular tasks. Although lab group size frequently is determined by the amount of equipment available, we suggest lab groups of two to three students to allow each student to play an active role. Assign responsibilities to each group member (e.g., recorder, materials management, quality control, cleanup, and safety officer). Rotate responsibilities and groups frequently.

Organize the lab equipment. Organizing lab materials into trays helps in the distribution of materials. Provide supply tables with easy access. Label all chemicals and unique items. Arrange desks or tables for adequate workspace. Reinforce safety issues particular to each lab activity with written instructions at both the lab table and on lab equipment trays.

Don't grade everything. Synergized lessons help you identify the points of assessment. Score only the activities or questions that give you a clear picture of student progress. Remember, evaluating student performance always involves scoring samples of student work, not everything the student does.

Plan for behavior issues. Work with your students to develop a set of classroom rules and the consequences if the rules are not followed. Temper the rules based on your situation. Make sure you let students know that parents must be contacted when rules are broken and that other staff may become involved (e.g., resource teacher and school administrator), if warranted.

Bring closure to your lab activities. Identify an appropriate time to bring closure to the student learning for the day, assign homework, and clean up. This is a great opportunity to validate student learning – What did we do today? What did we learn? Why is this important?

Working with Diverse Learners

Providing optimum experiences for all students within your classroom is a challenging endeavor. Language issues, group dynamics, safety concerns, and intellectual maturity can impact the success of students. *SMSC* lessons address these factors in the teacher notes by including pre-lesson information, options within investigations, reading strategies, and adaptable assessments.

Collaborative Teaming

Synergized lessons require students to work in teams. This partnership, with a variety of skill levels and personalities, will determine the success of the lesson. Consider these strategies when organizing lab teams.

- Purposely assign students to teams. Identify issues with specific students: allergies, learning disabilities, physical disabilities, personal issues, and behavioral issues. Develop a plan so that all students can be successful in a lab setting.

- Assign each student an active role such as recorder, materials manager, timer, and safety officer.
- Let students know grading expectations. Will assignments be scored with a rubric? Will everyone in the group get the same grade? Will only one notebook or piece of work be collected and scored? If collecting work from everyone in the group, can any work be exactly alike?
- Keep lab groups small. The number of students per group should be as small as possible given the limitations on equipment. When teams are smaller, all students will be more engaged.
- Minimize the impact of the disruptive students. Separate disruptive students into different groups, monitor their behavior, ask resource staff for assistance, and/or provide an alternative assignment. One student's disruptive behavior can ruin the learning experience for the entire class.

Customizing Lessons

Students in a science class are likely to have multi-level skills, abilities, and talents. Assignments should be designed to challenge, but not to overwhelm the student. *SMSC* lessons can be differentiated for inquiry, literacy, mathematics, and unique skills, talents, and abilities.

Differentiation for Inquiry Investigations

The National Research Council's book, *Inquiry and the National Science Education Standards* states that inquiry can be structured to meet varying learning outcomes; such as, learning concepts, acquiring inquiry abilities, or developing understandings about inquiry. They also note that inquiry can vary along a continuum from structured and teacher-guided to open-ended and student-directed. (NRC, 2000)

From our experiences, English language learners and special needs students benefit from more structured, guided inquiry lessons. On the other hand, highly motivated and gifted students benefit from self-directed learning. The NRC states, "Guided inquiry can best focus learning on the development of particular science concepts. More open inquiry will afford the best opportunities for cognitive development and scientific reasoning (NRC, 2000, p. 30)."

In structured inquiries, students follow a prescribed inquiry format that includes: focus questions, predictions, observations and data collection, and stating claims and evidence. In guided inquiries, students may have more freedom in the selection of variables but also follow a prescribed format. In open inquiry, students are given the freedom to develop or diverge within given parameters. Synergized investigations are designed as adaptable guided inquiry lessons. Teachers can add more structure or reduce structure as needed.

Differentiation for Literacy

Students who do not read at grade level and those who struggle with both speaking and writing in English can succeed in a science classroom. There are opportunities in *SMSC* lessons for teachers to modify and/or adapt lessons to a variety of literacy levels.

- Provide both verbal and written instructions for investigations and activities.

- Model and demonstrate lab techniques.
- Illustrate concepts with pictures, diagrams, and other media.
- Provide graphic organizers for students to use to record observations and conclusions.
- Provide formats for data tables and graphs (many are already provided in the lessons).
- Identify key vocabulary and use vocabulary building strategies like word splashes, three-column vocabulary charts, word wall activities, and vocabulary cards.
- Allow students to use graphics and/or pictures to explain concepts.
- Alternate the way students are assessed by using oral questioning and demonstrations of learning as well as modifications of written work.

Differentiation for Mathematics

Students who struggle with mathematics concepts and their application may need special attention. There are opportunities in *SMSC* lessons for teachers to modify and/or adapt lessons to a variety of mathematical abilities.

- Review mathematical concepts applied in science lessons.
- Align the instructional methods used to teach mathematics with science instruction.
- Scaffold mathematical computations to include a number of sample problems.
- Practice mathematical skills in context (Many students get their mathematic "aha moment" when they apply the mathematics in a real situation).
- Offer tutor time for struggling students.

Differentiation for Unique Skills, Talents, and Abilities

When lessons are modified to take into account the unique skills, talents, and/or abilities of students, the students will perform at higher levels and take more ownership in their learning. To adapt lessons for unique learners, consider the following:

- Provide opportunities to incorporate artwork, computer skills, communication skills, and leadership abilities in *SMSC* lessons.
- Allow students to choose different ways to demonstrate their learning: posters, oral presentations, multi-media products, writing formats (e.g., poetry, essay, and children's book), and dramatic performances (e.g., songs, plays, and speeches).

Working with Support Staff, Parents, and Classroom Volunteers

Parents and/or guardians are an integral part of the educational process for every student. They should have regular communication with their student's teachers and the school. Parents can provide the support needed at home as well as valuable insight into their student's needs and abilities.

The success of the synergized classroom is greatly enhanced by the inclusion of school support staff, administrators, and classroom volunteers. Special education staff can work closely with identified students and help modify assignments and expectations. Support staff as well as parent volunteers can provide one-on-one instruction for struggling students. Classroom volunteers can assist with materials management and classroom organization. School administrators can assist with issues that impact student learning and safety.

SAFETY IN THE SCIENCE CLASSROOM

Although all *SMSC* lessons are designed with safety and safe materials in mind, accidents can happen. Use common sense when preparing, implementing, and cleaning up after lessons. You are professionally responsible for the safety of your students. Synergized science lessons incorporate tried and true science experiments and activities. The lessons selected were chosen because they use readily available safe, materials.

It is important to set the stage for safe laboratory behaviors at the beginning of the course. In Matter's Phases, safety habits used throughout the chemistry unit are addressed. The following general safety tips are also recommended.

• *Always practice the lesson procedures before you use the lesson with students.* Doing lessons ahead of time allows you to note any potential problems and safety issues in your unique setting and significantly reduces the probability of student accidents.

• *Distribute lab safety rules to all students.* Discuss the rules. Display a large version of the rules in your classroom. Require students and parents to sign safety contracts that clearly state the rules.

• *Formulate a plan to deal with safety violations.* Let students know the consequences for safety violations. Do not tolerate horseplay. Make the consequence fit the offense.

• *Demonstrate safety equipment.* Identify for students the locations and use of the fire extinguisher, fire blanket, safety goggles, eye wash station(s), and any other safety equipment in the classroom. Demonstrate how and when to use personal protection equipment (PPE) such as safety goggles, aprons, and gloves.

• *Establish a Procedure to Handle Safety Issues.* Share with students specific protocols to deal with chemical spills, broken glassware, and student injuries. Remind them that all accidents need to be reported to the teacher immediately.

• *Read and organize Materials Safety Data Sheets (MSDS).* Chemicals purchased from a supply house will have a Materials Safety Data Sheet (MSDS). These should be read and placed in an easy to access file before the chemicals are used with students.

• *Safely dispose of any waste materials generated in your classroom.*

• *Obtain a copy of the American Chemical Society Science Safety Guidelines.* The American Chemical Society (ACS) provides valuable safety information for teachers. You can download a copy from the ACS website: www.acs.org.

References

Marzano, Robert J., D. Pickering, and J. Polluck. 2000. *Classroom instruction that works.* Alexandria, VA: Association for Supervision and Curriculum Development.

National Research Council (NRC). 1996. *National science education standards.* Washington: DC: National Academy Press.

National Research Council (NRC). 2001. *Inquiry and the national science education standards.* Washington: DC: National Academy Press.

Doran, Rodney. F. Chan, P. Tamir, and C. Lenhardt. 2002. *Science educator's guide to laboratory assessment.* Arlington, VA: NSTA Press.

Lesson Set One:
Matter's Phases

OVERVIEW

<u>Lesson Set Concepts</u>

1.1 **Matter** is anything that has mass (as measured by weight) and takes up space.

1.2 A basic property of matter is its **state**. The three basic states are **solid, liquid, and gas (vapor)**.

1.3 Matter changes state or phase due to the conditions of its surroundings. The temperature and pressure of the surroundings determine phase. For example, under standard atmospheric pressure, water is a solid (ice) at or below 0°C (32°F), a liquid between 0°C and 100°C (32°F and 212°F), and a gas (water vapor) when the temperature is over 100°C (212°F).

1.4 In general, solids have a fixed shape and volume. Liquids take the shape of their container and are pourable. Gases have no fixed shape and are compressible.

1.5 Each state of matter can be identified by its characteristic properties. **Solid**: crystalline shape, solubility, and melting point. **Liquid**: ability to dissolve or not dissolve solids and/or gases, freezing point, and boiling point. **Gas:** volume is observably affected by even moderate temperature and pressure changes.

1.6 Matter is made of smaller units called **atoms.** Atoms can exist singly or combine with each other or other types of atoms to form **molecules**.

1.7 Atoms and molecules have **kinetic energy** or energy of motion. Atoms and molecules of solids have the least amount of **kinetic energy**. Atoms and molecules of gases have the most amount of **kinetic energy**.

1.8 When **heat energy** is transferred to a solid, the atoms or molecules gain kinetic energy. As more energy is added, the solid's atoms or molecules start to spin, rotate, and move. As more energy is added, the solid melts and changes to its liquid phase. As more energy is added to the liquid, the liquid evaporates into its gas or vapor phase.

1.9 Due to their **kinetic energy**, atoms or molecules of solids vibrate but don't move about, rotate, or spin. Atoms and molecules of liquids vibrate, move about, rotate, and spin. Atoms and molecules of gases vibrate, rotate, spin, and spread apart moving much faster than in their liquid phase.

1.10 **Strong attractive forces** exist between individual atoms and molecules. These forces hold atoms and molecules of solids and liquids together. As liquids phase into their gaseous state, the attractive forces no longer hold the atoms or molecules together due to their increased kinetic energy. Thus, atoms and molecules of gases spread out as they bombard each other.

1.11 **Diffusion** occurs as atoms and molecules of one kind of matter randomly mix with atoms and molecules of another kind of matter. Atoms and molecules randomly move from an area of higher concentration to lower concentration.

1.12 Phase changes are processes that describe changes in state. Phase changes are: **melting** (solid to liquid), **freezing** (liquid to solid), **evaporation** (liquid to gas), **condensation** (gas to liquid), **sublimation** (solid to gas), and **deposition** (gas to solid).

Key Vocabulary: matter, state of matter, solid, liquid, gas, vapor, atom, molecule, phase change, melting, freezing, evaporation, condensation, sublimation, deposition, kinetic energy, attractive forces, diffusion

Lab Equipment and Supplies: Matter's Phases		
General Science Equipment & Supplies	**Grocery Store: Paper, Plastics, Specialty**	**Grocery/Box Stores/Pharmacy: Chemical Supplies**
Per Class: 2-3 Electronic Balances Hot Plate Tongs Six-Well Plates Graduated Cylinders: 25, 50, and 100 mL Petri plates Beakers, 50 mL Pipettes (eyedroppers) Per Student: Science Notebooks Safety Goggles Calculators Metric rulers Graph Paper Per Group: Hand Magnifiers Special Order Item: Micro-scale Vacuum Apparatus (See Teacher Notes Lesson 16 for ordering information)	Ice trays for ice Paper Towels Plastic Wrap Punch cups, plastic 8 oz. Bottles, plastic 24 oz. Food storage containers, 5 Styrofoam balls-small Wax paper Toothpicks Cotton swabs Crayons, red/blue Spoons, plastic Masking Tape Rubber Bands Index Cards Construction Paper Brown Paper (5cm x 5cm) Craft sticks Balloons String Cans (empty aluminum) Straight pins Dish pan/plastic storage box Scissors Pencil Butcher/Poster paper Markers	Salt, Rock Salt, Table Epsom Salt Borax Sugar Water, tap or distilled Mineral oil Rubbing Alcohol Vinegar, white Food Coloring Set Bubble solution Liquid dish detergent Baking Soda Marshmallows Shaving cream Food Extracts (8-10 types)

	Lesson Set Guide: Matter's Phases		
Lesson		**Title**	**What Students Do**
1 Intro & Assess		Everyday Chemicals	Students list the chemicals they use everyday, describe the chemicals with words or phrases, and then pair-share to see if their partner can guess the chemical. Students combine their everyday chemical lists and classify the chemicals in different ways. They also classify the chemicals as solids, liquids or gases.
2 Intro Safety		Chemical Safety and Safety Rules	Students read and discuss the chemical safety rules.
3 Invest		A Solid Identification	Students determine the identity of an unknown white solid by comparing the unknown to four other white solids. Appearance: Students compare solids by observing their size, color, texture, sheen, transparency, and other appearance characteristics. Solubility in Water: Students determine how soluble each solid is in water. Crystallization: Students make saturated solutions of the solids, allow the solids to crystallize overnight, and then compare their crystals.
4 Read		A Solid Start	Students read about the properties of solids using a word splash to connect words and ideas.
5 Demo		Melting and Freezing	Students observe and discuss as the teacher uses a simple model to illustrate the molecular motion of solids and liquids during melting and freezing.
6 Invest		Comparing Water's Properties to Other Liquids	Students compare the behavior of water to vinegar, mineral oil, and rubbing alcohol. •Behavior of Drops: Students place drops of each liquid on plastic wrap and wax paper and compare the drops' behavior. •Paper Absorption: Students place drops of each liquid on brown paper and paper towels and compare how the liquids are absorbed into the paper. •Solubility of Sugar: Students determine how each liquid dissolves sugar. •Mixing Liquids: Students determine how well the liquids mix with each other.
7 Read		Phasing into Liquids	Students read about the properties of liquids.
8 Invest		Observing Kinetic Energy of Liquids	Students indirectly observe the kinetic behavior of food coloring mixed with water at different temperatures. Students are introduced to the concept of diffusion.
9 Invest		Explore Some More – Evaporation Inquiry	The students design and conduct their own investigation to determine the factors that increase or decrease the speed of the evaporation rate of water.

Lesson Set Guide: Matter's Phases Continued

Lesson	Title	What Students Do
10 Invest	Condensation	Students relate the effect of temperature changes to condensation.
11 Invest	Air Matters	Students provide evidence that air is made of matter by designing a poster illustrating how air takes up space and has mass.
12 Invest	Gases and Temperature Change	Students observe how a change in temperature affects the volume of a gas.
13 Read	Comparing Matter's Phases	Students read about the properties of gases and compare and contrast solids, liquids, and gases.
14 Invest	Diffusion-Do You Have a Super Nose?	Students learn about diffusion by identifying the scents of a variety of different food extracts that have been placed in balloons.
15 Read	Diffusion	Students read about diffusion, identify the five most important words, and then define the words in their own words.
16 Invest	Gases and Air Pressure	Students observe how differences in air pressure affect objects and phase changes.
17 Read	Boiling Hot or Boiling Cold?	Students read about air pressure and how it affects evaporation, condensation, and the boiling of liquids and then write a summary of their reading by connecting words and ideas.
18 Demo	Can Crush	Students observe a can crushing demonstration and make observations. After observing the demonstration repeated a number of times, students relate their observations to the processes of evaporation, condensation, and boiling.
19 Graph	Graphing Phase Changes	Students construct graphs depicting the temperature change of liquids over time. Students interpret data and graphs to determine the boiling points and freezing points of three liquids.
20 Review	Reviewing Matter's Phases	Students read about sublimation and deposition. They work in pairs to complete a word sort. Students make a concept map showing the relationship of matter's phases. Students use vocabulary in a novel way. Students revisit their everyday chemical lists to review classification and answer additional questions.
21 Assess	Matter's Phases Post Test	Students participate in a formal assessment of their learning. Students select from a variety of phase terms to write sentences that connect terms and meanings. Students answer multiple-choice questions.

5

		Before and During the Lesson		End of Lesson
Assessment Guide: Matter's Phases				
Lesson	**Title**	**Uncovers Student Ideas**	**Checks for new understandings**	**Evaluates learning**
1 Intro & Assess	Everyday Chemicals	•On Your Own & Pair-Share •Notebook Entry	•Team Work •Notebook Entry	See Lesson 20: Revisit Everyday Chemicals
2 Safety	Chemical Safety	Class Discussion	Daily Monitoring of Safe Practices	Teacher Designed Performance Assessments, Quizzes, Tests
3 Invest	A Solid Identification	•Focus Questions •Reflections	Claims & Evidence (C&E) Science Roundtable Discussion	New Understandings (NU): 1
4 Read	A Solid Start	Word Splash	Word Splash	Word Splash Summary
5 Demo	Melting & Freezing	Concept Modeling	Concept Modeling	Notebook Entry
6 Invest	Comparing Water's Properties to Other Liquids	Focus Questions	Part 1. C&E NU Part 2. C&E NU Part 3. C&E NU Part 4. C&E: 1,2	Part 4. C&E: 3 Part 4. NU: 1,2,3
7 Read	Read: Phasing into Liquids	Lesson Six	QAR	QAR
8 Invest	Observing Kinetic Energy of Liquids	Lesson Seven Hypothesis Reflections	Notebook Entries NU: 1, 2	C&E: 1,2
9 Invest	Explore Some More – Evaporation Inquiry	Focus Question Reflections	C&E Notebook Entry: Observations	NU: 1,2,3
10 Invest	Condensation	Focus Question Reflections	NU: 1,2,3	C&E
11 Invest	Air Matters	Focus Question Reflections	List & Select NU: 3,4	NU: 1,2
12 Invest	Gases and Temperature Change	Focus Question Reflections	Observations C&E: 1,2,3,4,5	NU: 1,2

		Before and During the Lesson		End of Lesson
Lesson	Title	Uncovers Student Ideas	Checks for new understandings	Evaluates learning
13 Read	Comparing Matter's Phases	Feature Analysis	Feature Analysis	Feature Analysis
14 Invest	Diffusion - Do You Have a Super Nose?	Focus Question Reflections	C&E; NU: 1, 2 Reflections	NU: 3
15 Read	Diffusion	Lesson 14	Five Most Important Words	Notebook Entry: Definition of Diffusion
16 Invest	Gases and Air Pressure	Focus Question	Procedure Observations NU: 1, 2, 3	C&E: 1, 2
17 Read	Boiling Hot or Boiling Cold?	Anticipation Guide I & II	Anticipation Guide III & IV	Anticipation Guide V
18 Demo	Can Crush	Teacher Demo Focus Question Initial Observations	More Observations C&E: 2 NU: 2, 3 Reflections	C&E: 1 NU: 1
19 Graph	Graphing Phase Changes	Focus Question Reflections	Notebook Entry: Graphs BP: NU: 2, 3 ,4 ,5 FP: NU: 1, 2, 3	BP: DI: 1,2,3,4,5 NU: 1, 3, 5, 6 FP: DI: 1, 2, 3, 4, 5, 6 NU: 1
20 Review	Reviewing Matter's Phases	Word Sort First Word-Last Word FW-LW: Q1 Everyday Chemicals	Word Sort FW-LW: Q2 Notebook Entry: Revisions	Notebook Entry: Concept Map FW-LW: Q3
21 Assess	Matter's Phases Post Test			Part 1: Multiple Choice Part 2: Match & Write

Assessment Guide: Matter's Phases Continued

Lesson One: Everyday Chemicals
Student Introduction and Pre-Assessment

Chemistry is the study of everything everywhere! Chemists seek answers to basic questions about matter like: What is it? What's it made of? How can it be described? Does it change with time? Is it man-made or found in nature? And, on and on... A good way to begin your study of chemistry is to find out what you already know. And, a fun way to find out what you already know is to play a game.

On Your Own
1. In your notebook, list "chemicals" you use everyday? (2 minutes)
2. Pick one of your "everyday" chemicals, and describe it in as many ways as you can using single words or short phrases. (2 minutes)

Pair-Share
1. Without telling your partner the name of your everyday chemical, read just your descriptive words or phrases and see if your partner can guess your everyday chemical. (4 minutes)
2. After you have both described your everyday chemical, compare your chemical lists and see how many chemicals you named in common. (4 minutes)

Team Work
Because there are so many different chemicals, chemists have found ways to group or classify chemicals for study and reference. Next, you will group your everyday chemicals.
1. Using a large sheet of paper, combine your list of everyday chemicals with the lists of at least one other group. If you don't know about a chemical on the list, ask the person who listed it to explain it to you.
2. Determine how many ways the chemicals on your list could be placed into smaller groups. For example, you could group them by where they are found (the kitchen, car, and yard). How many other ways can you group the chemicals on your list?

Grouping Matter by Phases
Chemists group all of the substances into a larger category called **matter. Matter is anything that has mass (as measured by weight) and takes up space.** Mass is the amount of stuff in a substance. Chemists use the word "mass" because weight measurements change depending on the pull of gravity. For example, on the Moon, astronauts weigh less than on Earth, but still have the same mass.
One way chemists classify matter is by **state**. The three basic states of matter are: solid, liquid, and gas (vapor). For example, water's solid state is ice, liquid state is water, and gaseous state is water vapor.

More Team Work
From the combined chemical list, write an **S** by the solids, **L** by the liquids, and **G** by the gases. Have your teacher check your work.

On Your Own: Make three columns titled solid, liquid, and gas. Select five substances for each column.

Teacher Notes: Everyday Chemicals
Lesson One: Student Introduction & Pre-Assessment

Rationale: This activity sets the stage for the lesson set. Designed as a pre-assessment, it prompts student discussion of everyday chemicals as well as provides insight into what students call chemicals and how they classify chemicals. Students list, classify, and justify categories to activate prior knowledge and generate questions.

Materials & Preparation: Student copies of *Student Lesson One: Everyday Chemicals*, butcher paper, tape, and markers. The student page does not need to be copied for every student. It can be projected on a screen or used as a teacher facilitation guide.

Teacher's Role: Your role is to facilitate a collaborative learning environment that encourages student-to-student discussion about chemicals and how chemicals are described and grouped. The terms **solid, liquid, and gas or vapor** are purposely not defined so that you can uncover student ideas as they classify their everyday chemicals. The term **matter** is defined and the **concepts of mass and weight** are introduced. In future lessons, students will be asked to find the mass of substances rather than the weight. Thus, it is suggested that a more detailed explanation be postponed to a time when students are massing chemicals. Group classification lists and student questions should be posted for future reference. During and/or after the lesson, current student understandings, preconceptions, and questions should be noted for inclusion into future lessons.

Background Information: This classification activity pre-assesses the students' current understanding of the three basic states of matter: solid, liquid, and gas (vapor). Even though students may have experienced elementary science lessons on solids, liquids and gases, we've found that middle school students need to revisit the concepts. The definition of matter is introduced as anything that has mass (as measured by weight) and takes up space. Mass is defined as the amount of stuff in a substance, and weight is defined as a measure of mass.

Teaching Tips

1. Because this is a pre-assessment, your role is to generate student discussion rather than provide direct instruction about solids, liquids, and gases (vapors).

2. Take notes on student ideas and preconceptions. Use your notes to remind you to highlight specific concepts and ideas about solids, liquids and gases in future lessons.

Teacher Notes: Chemical Safety and Safety Rules
Lesson Two: Student Safety

Rationale: Instructing, monitoring, and enforcing chemical safety is part of every science lesson. If you should have an accident in your classroom, you want to make sure that you did everything a reasonable teacher would do to provide a safe learning environment for student work. Before you begin any lesson, you must know the safety of the chemicals and any procedural cautions. The chemicals and materials in the lessons that follow are specifically designed for middle school because they use safe, common chemicals that can be locally obtained in supermarkets and large discount stores. The lessons also pose no disposal hazards. However, accidents can still happen if safety rules are not followed, monitored, and enforced.

Reading Strategy: Shared Reading – Everybody Read To... (ERT). See Literacy Resource Section.

Materials and Preparation: Basic Chemical Safety Rules – Student Copy and Teacher Copy (follow teacher notes). The student copy of the rules is kept in their science notebook. The teacher copy is kept in the teacher's safety file.

Before Class: Read the **Basic Chemical Safety Rules** provided and determine if additional rules apply in your situation and add them to the list.

Teacher's Role: The teacher is always responsible for providing the safest possible classroom environment. Make sure you have read the list before reading the rules with your students.

Read the **Basic Chemical Safety Rules** with your students using the shared reading format known as Everybody Read To…(ERT) outlined below. In ERT, "to find out" means to look in the reading for the response and "to figure out" calls for students to infer and or bring their own background experiences to the reading.

1. Ask the students to read *to find* the rule that deals with chemical labeling and *to figure out* why the rule is necessary. Tell them to raise their hand when they have an answer. Wait until most of the class has their hands raised. Call on students to give the answers. Finally, call on another student to read the rule to the class.

> **Possible Student Responses**
> **First Student**: Rule one deals with chemical labeling. Lab chemicals can be mixed with other substances. They are used from year to year. They could make you sick. There could be a mistake in the labeling….
> **Second Student:** Rule one states: Never taste a lab chemical. Use only labeled chemicals approved by your teacher. Class chemicals are for lab purposes only and may be impure or contaminated with other substances. Many chemicals look alike. Use only properly labeled chemicals.

2. Repeat the above process for all the rules by asking questions that require students to both *find out* the rule and *figure out* why a rule is necessary. You should not ask about the rules in the order on the sheet. And, some questions may apply to more than one rule. For example, rules four and five both apply to working with chemicals in the lab.

3. After all rules have been read and discussed, have each student write their names and date on both sets of rules. One set of rules is kept in the student's science notebook for reference, and the other set of rules is signed by parents/guardians and filed in the teacher's safety file. If a student is absent, go over the rules individually as soon as the student returns.

4. Post a set of the safety rules in your classroom in a conspicuous place. In your lesson planner, note and highlight all safety lessons as evidence of your commitment to safe practices.

You must always model safe behaviors. If safety glasses need to be worn, you need to wear safety glasses. You must never leave a class during a lesson. Hands-on activities or demonstrations must be performed prior to use in the classroom to make sure you can perform them safely and to determine if additional safety cautions apply. No matter how minor, accidents must be reported to school administration and parents/guardians notified. Consequences must be made clear for students who do not follow the rules.

Background Information: The **Basic Chemical Safety Rules** are to be read and acknowledged by students, parents/guardians, and teachers. The rules represent critical areas of laboratory and chemical safety for middle level students; however, *the Basic Chemical Safety Rules do not cover every aspect of student safety in detail as each school and classroom has unique needs. These rules do not replace rules that are required by a school or district that are already posted and being enforced.* Before each investigation, appropriate rules need to be reiterated and any additional rules and cautions added and taught.

Teaching Tips
• **Incorporate additional safety lessons in future lessons**. Once the rules are formally introduced, other safety lessons can and should be added to the curriculum. For example, student groups can be assigned a rule to illustrate through a poster or safety icon. The posters and/or icons can be displayed in the classroom. Questions about safety should be a part of assessments. Students can act out safety situations and an appropriate response.

• **Formulate a plan to deal with safety violations**. Let students know the consequences for safety violations. Do not tolerate horseplay. Allow students to help you develop consequences for safety violations. Make sure the consequence fits the offense. For example, if a student is lax in wearing safety glasses or goggles, they should have to write a statement explaining why it is critical to wear safety glasses or goggles and read it to the class. And, they should not be permitted to perform any lab activity without safety glasses or goggles.

- **Consider chemical disposal and clean up.** Train students to dispose of chemicals properly. Make sure students know what to do in case of spills. Make sure they know that they must report any spills or broken equipment to you before trying to clean it up or repair it. For example, broken glass can be especially dangerous and should be cleaned up by the teacher or other staff member.

- **Demonstrate safety equipment.** Identify the locations and use of the fire extinguisher, fire blanket, safety goggles, eye wash station(s), and any other safety equipment in the classroom.

- **Monitor student behavior for every lesson.** Immediately deal with any safety issues or student behavior related to unsafe practices.

- **Personal Protection Equipment (PPE).** Demonstrate how and when to use personal protection equipment such as safety glasses, goggles, aprons, and gloves.

- **Formally assess and evaluate safety and lab performance.** Have the students participate in the design of a safe-practices lab checklist. Use this assessment checklist, to evaluate lab behavior on a regular basis. Add questions related to safe lab procedures on written quizzes and tests.

Lesson Two: Basic Chemical Safety Rules
Student Copy with Student and Teacher Signatures

Chemists avoid accidents by knowing how to investigate safely with chemicals. They also make sure that other chemists are working safely. Listed below are basic rules for working safely in the classroom science lab. Before you begin, your teacher will go over the rules. Your teacher may also add additional rules. When in doubt about a chemical or procedure, always check with your teacher. And, never perform any procedure unless it is teacher approved.

As your teacher reviews the rules below, place your initials after each rule to show that you understand and will follow the rule. Have your teacher initial the list when you are done. Add this sheet to your science notebook.

1. **Never taste a lab chemical. Use only labeled chemicals approved by your teacher.** Class chemicals are for lab purposes only and may be impure or contaminated with other substances. Many chemicals look alike. Use only properly labeled chemicals.
2. **Wear appropriate personal protective equipment. This includes chemical splash goggles, lab aprons, hair clips, closed-toe footwear, and other equipment as needed.**
    *   Goggles protect your eyes from chemical spills. They also prevent you from rubbing your eyes and transferring a chemical from your hands to your eyes. If you should get something in your eye, immediately follow proper procedures to wash your eyes.
    *   Aprons or smocks can be worn to protect clothing.
    *   Wear shoes that cover toes as chemicals can spill on feet exposed with sandals or flip-flops.
    *   Keep hair out of experiments by using clips, bands, or tiebacks.
3. **No rough housing or horseplay in the classroom lab.** Avoid accidents or injuries by always behaving appropriately.
4. **Properly dispose of all chemicals.** Your teacher will provide directions for the safe disposal of the chemicals and the materials you use.
5. **Report all spills to your teacher.** Follow your teacher's directions for cleaning up spills.
6. **Report all glass breakage. Do not throw broken glass in the trash.** Never pick up broken glass with your fingers. Your teacher will provide clean up materials and show you how to dispose of broken glass.
7. **Read and understand all procedures and directions before beginning lab work.** Many accidents can be avoided if everyone follows directions and obeys the safety rules.
8. **Know where all class safety equipment is located.**
9. **Immediately report any accident or injury to your teacher, no matter how minor it may seem.**
10. **Always show respect to others and their work.**
11. **Follow all other rules established by your teacher and your school.** List any additional safety rules established by your teacher or attach the list to this page.

*Date:*_____

*Student Signature:*_____

*Teacher Signature:*_____

Lesson Two: Basic Chemical Safety Rules
Teacher Copy with Student and Parent Signatures

Dear Parents/Guardians: We are beginning a chemistry unit. Although we will be using safe, common chemicals, it is important that students learn proper chemical safety. Please sign this sheet acknowledging that your student has read and explained the rules to you.

Chemists avoid accidents by knowing how to investigate safely with chemicals. They also make sure that other chemists are working safely. Listed below are basic rules for working safely in the classroom science lab. Before you begin, your teacher will go over the rules. Your teacher may also add additional rules. When in doubt about a chemical or procedure, always check with your teacher. And, never perform any procedure unless it is teacher approved.

1. **Never taste a lab chemical. Use only labeled chemicals approved by your teacher.** Class chemicals are for lab purposes only and may be impure or contaminated with other substances. Many chemicals look alike. Use only properly labeled chemicals.
2. **Wear appropriate personal protective equipment. This includes chemical splash goggles, lab aprons, hair clips, closed-toe footwear, and other equipment as needed.**
   - Goggles protect your eyes from chemical spills. They also prevent you from rubbing your eyes and transferring a chemical from your hands to your eyes. If you should get something in your eye, immediately follow proper procedures to wash your eyes.
   - Aprons or smocks can be worn to protect clothing.
   - Wear shoes that cover toes as chemicals can spill on feet exposed with sandals or flip-flops.
   - Keep hair out of experiments by using clips, bands, or tiebacks.
3. **No rough housing or horseplay in the classroom lab.** Avoid accidents or injuries by always behaving appropriately.
4. **Properly dispose of all chemicals.** Your teacher will provide directions for the safe disposal of the chemicals and the materials you use.
5. **Report all spills to your teacher.** Follow your teacher's directions for cleaning up spills.
6. **Report all glass breakage. Do not throw broken glass in the trash.** Never pick up broken glass with your fingers. Your teacher will provide clean up materials and show you how to dispose of broken glass.
7. **Read and understand all procedures and directions before beginning lab work.** Many accidents can be avoided if everyone follows directions and obeys the safety rules.
8. **Know where all class safety equipment is located.**
9. **Immediately report any accident or injury to your teacher, no matter how minor it may seem.**
10. **Always show respect to others and their work.**
11. **Follow all other rules established by your teacher and your school.** List any additional safety rules established by your teacher or attach the list to this page.

*Date:*_____

*Student Signature:*_____

*Parent/Guardian Signature:*_____

Lesson Three: A Solid Identification
Student Investigation

Introduction: Part of a chemist's work is to identify unknown chemicals. There are many reasons to identify a chemical. For example, forensic chemists try to match chemicals found at a crime scene to substances found on a suspect's clothing or shoes. Environmental chemists must identify substances in drinking water to make sure the water is safe. And, all scientists must know the chemicals they use so that the chemicals can be used safely and disposed of properly.

Chemists begin their identification of an unknown substance by observing and testing the substance. Chemists also make comparisons to known substances. The more a chemist learns about an unknown substance, the better the possibility of a correct identification.

Problem: Your team will be given an unknown, white solid labeled "U" to identify. It is one of the four white solids labeled A, B, C, and D. Your job is to gather evidence to support your identification of solid U as solid A, B, C, or D. Although you don't know the identities of A, B, C, or D, it is important to know that they are common, everyday chemicals.

Focus Question: What is the identity of the solid labeled U?

Prediction: If you think you know the identification of solid U, make a prediction and tell how you know. Or, tell why you would rather not predict at this time.

SAFETY: NEVER taste any lab chemicals, even if you know what they are. Chemicals used in the lab are often reused, impure, and unsanitary for human consumption.

Pre-lab Preparation: Before you begin testing solid U, get the materials listed below and follow the procedure for labeling your test chemicals.

General Materials (2-3 students per group): Safety goggles, paper towels, masking tape, plastic spoons, hand held magnifier, samples of U (Unknown), A, B, C, and D; 6-well plate or 5 small cups

Labeling Procedure
1. Using masking tape, label five sections of the well plate: U for unknown, A, B, C, and D for the other four substances.
2. Place a spoonful of each solid in the correctly labeled section. You will use these solids in the procedures that follow.

A Solid Identification

Part 1. Appearance: You will compare the appearance of U (unknown white solid) to A, B, C, and D. You will compare particle size, shape, color, texture, sheen, and the ability of light to pass through the solid. You may add additional observations, if desired.

Focus Question: How does the appearance of U compare to solids A, B, C, and D?

Materials (2-3 students per group): General materials

Procedure 1. Appearance
1. Copy the chart below in your science notebook.
2. Observe small amounts of each solid with your magnifier. Note: Transparent substances allow light to pass through. Opaque substances do not allow light to pass through.
3. Record your observations in your chart.

Comparing Observable Features of Unknown: U to A, B, C, and D					
Feature	Unknown U	A	B	C	D
Particle Size					
Color					
Texture					
Sheen: Shiny vs. Dull					
Transparent vs. Opaque					
Other					

Claims & Evidence: Can you claim solid U (unknown) is A, B, C, or D based on your observations? If you can, state the identity of U (A, B, C, or D) and use evidence to support your claim. If not, tell why not.

A Solid Identification
Part 2. Solubility in Water: How Well Does a Solid Dissolve?
A solid is soluble in a liquid if it dissolves. When a solid substance dissolves, its particles spread evenly throughout the liquid. Even though the solid may seem to disappear, it is still there. The particles are just too small to be seen. All substances do not dissolve the same in a liquid. Some substances will dissolve more completely than others or some will not dissolve at all.

Focus Question: How does the solubility of U in water compare to the solubility of A, B, C, and D in water?

Materials: In addition to the general materials you will need: Safety goggles, water, 100 mL graduated cylinder, plastic punch cups, craft or stirring sticks

Procedure 2. Solubility in Water
1. Copy the chart below in your science notebook.
2. Measure 100 mL of water and add it to a plastic punch cup.
3. Place a level spoonful of U into the water.
4. Stir and observe.
5. Repeat for each substance.
6. Record your observations in your chart. Note: Colorless and clear do not mean the same thing to a chemist. This is best explained by example. Pure water is both colorless and clear. Iced tea is not colorless (brown), but it can be clear. Orange juice is neither colorless nor clear.

Comparing the Solubility of U with A, B, C, and D in Water			
Substance	**Rankings** 1=did not dissolve 3=dissolved but cloudy or some solid remaining 5=all dissolved, liquid colorless and clear	**Describe the clarity of the water after stirring.**	**Other Observations**
Unknown			
A			
B			
C			
D			

Claims and Evidence: Can you claim solid U is A, B, C, or D based on how well the solids dissolved? If you can, state the identity of the unknown (A, B, C, or D) and use evidence to support your claim. If not, tell why not.

New Understandings
1. Do you think you would get the same results if you dissolved the substances in a liquid other than water?
2. Do you think the temperature of the water might affect your results?

A Solid Identification
Part 3. Crystal Formation
In this test, each of the five solids will be dissolved in water. The container will be left overnight for the water to evaporate. The solids will crystallize on the bottom of the container as the water evaporates.

Focus Question: How do the crystals of U compare to crystals A, B, C, and D?

Materials
In addition to the general materials you will need: 6-well plate or 5 Petri plates or other shallow containers for crystal evaporation, 50 mL small beaker, warm tap water

Procedure 3. Crystal Formation
1. Copy the Comparing Crystals Chart in your science notebook
2. Use masking tape and pen to label each evaporating container U, A, B, C and D. Also, label each container with your team name.
3. Fill the small beaker ½ full of warm tap water.
4. Add a spoonful of U to the warm water in the beaker and stir.
5. Pour a small amount of the U solution into the evaporating container so that it just covers the bottom.
6. Repeat steps two through five with A, B, C, and D.
7. Place the evaporating container(s) in a place designated by your teacher.
8. Leave overnight to evaporate.
9. The next day, use your magnifier to observe the solids left after evaporation.
10. Record your observations in your chart.

Comparing Crystals					
	U - Unknown	A	B	C	D
Crystal Drawing					
Other Notes					

Claims and Evidence

Can you claim solid U is A, B, C, or D based on how the solids formed crystals? If you can, state whether U is A, B, C, or D and use evidence to support your claim. If not, tell why not.

Science Round Table

Team Talk: Prepare to present your claims and evidence to the class. Use your testing information to make a convincing argument for the identity of chemical U.

Class Sharing: Share your findings with the class. As others present, compare and contrast their findings to yours. After each presentation, ask questions to clarify and learn more.

Summarizing Main Ideas: Use the questions below as a guide to summarize the class findings.
1. How did your results compare to other teams?
2. What team or teams presented the best claims based on their evidence?
3. How have your ideas changed as a result of the class discussion?

New Understandings
1. Based on the tests you have run and your previous knowledge of solids, what are the key properties of solids?
2. What other tests might you perform to learn more about the identity of the solids?
3. How might you improve the tests?

Reflections
1. How did your ideas compare to the others on your team and/or your class?
2. What additional questions do you have about chemistry as a result of performing this investigation?

Teacher Notes: A Solid Identification
Lesson Three – Student Investigation

Rationale: Students begin their study of matter's phases with an investigation designed to motivate learning and stimulate thinking. Students collect evidence to learn the identity of an unknown white substance. Their claims are challenged during class discussion as more evidence is presented. In the process, they learn that solids can be identified by the characteristic properties of solubility and crystal formation and that appearances can be deceiving. The investigation also serves as a before reading strategy for the next lesson.

Materials & Preparation: General Materials: Safety goggles, paper towels, hand held magnifiers, science notebook, pen, masking tape, labeled samples of U, A, B, C, and D, and plastic spoons; For Procedures: plastic punch cups, craft sticks or stirring sticks, graduated cylinder (100 mL), beaker (50 mL), five plastic food storage containers for stock quantities of U, A, B, C, and D

The Solids: U = rock salt, A = table salt, B = borax, C = sugar, and D = Epsom salt.
All substances were selected because they are safe, inexpensive in large amounts and found at the local supermarket. Borax is found in the laundry section and Epsom salt is found in the pharmacy section. Rock salt is used in pickling and canning, on sidewalks to melt ice, and in ice cream making. Kosher salt could be used if rock salt isn't available.

Place the solids in five different plastic food storage containers. Label both the containers and plastic spoons with the correct letter. Make sure that you don't give away the solids' identities by leaving their labeled store boxes where students can see them. Because rock salt is usually an unfamiliar form of salt to middle school students, it makes a good choice for the unknown.

Teacher's Role: Your role during the investigation is to facilitate lab work by managing the use of supplies, observing student work, and asking probing questions. As you monitor student work, continually assess student understanding as evidenced by their comments, questions, and notebook entries.

When students have completed Part 3, use the science round table strategy to guide students through the process of reconciling their findings with the findings of other groups. If this is the first time groups state claims and evidence, introduce or review appropriate class discussion behaviors, e.g., only one person speaks at a time, treat everyone with respect.

Determine the order of presentations by asking for volunteers or by a random method like throwing a die. During presentations, encourage clarifying questions but do not comment on the "right" or "wrongness" of their findings. After each presentation, thank each team for sharing their work. At the end of all presentations, summarize class findings. Allow students to change their initial claims to reflect any new evidence gained during the round table.

To preserve the "aha moment" of discovering the real identify of U, wait until all your classes have finished the investigation. Make clear that some physical properties like color and texture are not reliable identifiers. Lastly, show students the solids' packaging and read the label information. This is an excellent time to discuss package labeling.

Background Information: Students discover that physical appearances can be deceiving and that some properties are more useful in identification than others. Solubility or the amount of a substance (solute) that dissolves in a given amount of another substance (solvent) and crystalline shape are introduced as characteristic properties unique to each substance.

Part 1. Appearance: Rock salt and table salt look very different, although they are both sodium chloride. The differences in appearance result from pre-sale preparation. Rock salt is packaged in large chunks that have impurities making it not as white as the other solids. Table salt is finer grained and usually has additives to keep it from caking or to improve its nutritional value (iodine added). Borax is dispensed as a fine, white powder for use in the laundry. Table sugar has small, white crystals, and Epsom salt is chalky white with larger grains. Since all of the white solids appear different, students are intrigued and surprised to find out that two are actually different forms of sodium chloride or salt.

2. Solubility in Water: All substances dissolve in room temperature water ~ 25°C (77°F). The solubility of each substance in 100 mL of water at ~ 25°C is given below.

   Rock salt & table salt (both sodium chloride) = 35.9 g/100 cm^3
   Table sugar (sucrose) = 200 g/100 cm^3
   Borax (sodium tetraborate) = 5.8 g/100 cm^3
   Epsom salt (magnesium sulfate) = 25.5 g/100 cm^3

Rock salt dissolves slowly in water due to its larger grain size. The impurities cause the water to become cloudy and sometimes blackish-grey substances settle on the bottom. Table salt dissolves faster than rock salt, and the water remains cloudy due to the additives. Borax's fine particles dissolve quickly. Epsom salt also dissolves easily in water.

3. Crystal Formation: All of the substances form identifiable crystals; however, lab technique can alter results. By having all the groups make crystals, you should have examples to share with the groups that don't get identifiable crystals.

   Both rock salt and table salt form cubic crystals. Chemically, rock salt and table salt are sodium chloride or NaCl. Borax forms soft, many-sided, prismatic, colorless crystals. Chemically it is sodium tetraborate or $Na_2B_4O_5 \cdot 8H_2O$. The $8H_2O$ represents the water of hydration captured in borax crystals. As it dries out, the crystals lose water and become crumbly. It is used in many cleaning products to clean oily clothing and as a cleaning abrasive.

   Sugar forms rectangular and usually clear, colorless crystals. Chemically, it is called sucrose and consists of carbon, hydrogen, and oxygen ($C_{12}H_{22}O_{11}$). Epsom salt forms long, needle-like crystals. Chemically it is magnesium sulfate or $MgSO_4 \cdot 7H_2O$. Its solubility makes it a good choice to amend magnesium deficient soil. It is also used in footbaths to prevent skin wrinkling and reduce inflammation.

Lesson Four: A Solid Start – Word Splash
Student Reading

Introduction: Word Splashes build vocabulary and help you connect ideas in new and different ways.

Before Reading
1. Connect each word splashed on the page to at least one other word using a line.
2. On the line, describe the connection. For example, the words atom and molecule could be connected and described by saying that molecules are made of atoms.
3. Pair-share the reasons for your connections with a partner.

During Reading
1. All the splashed words are found in the reading. As you read, determine how the reading connects the words.
2. You may take notes during the reading.

After Reading
1. Make new connections for each word based on the reading.
2. Pair-share the new connections with a partner.
3. Finish by writing sentences that connect and include all the words in the splash. You may write a sentence that contains multiple splashed words.

Word Splash Master: A Solid Start

water

Atom

Matter

Property

lattice

kinetic

Solid

Motion

Liquid

Liquid

rotate spin-ids

vibrate

energy

ice

characteristic

molecule

phase

melting point

melting point

collide

crystal

Lesson Four: A Solid Start
Student Reading

What is a solid?

You learned that solids are not easily identified by color because many solids are white. Texture can also be misleading. The same solid can be chunky and sharp or ground into a fine, soft powder. A good way to identify solids is to find out how the solid acts when mixed with a liquid. But, an even better way is to observe the shape of a solid's crystal. This reading will help you learn more about the properties of solid matter.

If you could break down matter into the smallest unit that still is the substance, it will be an atom or a molecule. For example, the smallest unit of hydrogen that is hydrogen is a **hydrogen atom**. The smallest unit of oxygen that is oxygen is an **oxygen atom**. Two hydrogen atoms and one oxygen atom chemically combine to form water. Because water consists of more than one atom chemically combined, water is a molecule. The smallest unit of water that is water is a **water molecule.**

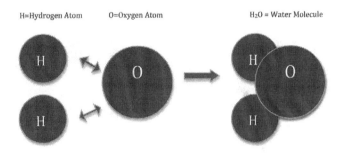

Atoms and molecules have energy of motion or **kinetic energy**. The kinetic energy of atoms and molecules differs based on their phase. Depending on temperature and pressure, the same substance can exist as a solid, liquid, or gas (vapor).

Solid state atoms or molecules are held together by strong attractive forces. The atoms or molecules are packed tightly together. In the solid phase, molecules do not move about because they don't have enough kinetic energy to overcome the attractive forces holding them together. However, they still have kinetic energy. Thus, their molecules are able to shake or vibrate.

The shape of a solid's crystal is a characteristic property of matter. Crystalline solids form crystal lattices that take on distinct geometric shapes; thus, crystalline shape is unique to each solid. For example, table salt always forms cubic crystals, and Epsom salt always forms needle-shaped crystals. In your investigation, all the white solids formed crystals. Some solids, like glass and tar, do not form crystals. These are called amorphous solids.

When a solid becomes a liquid, it **melts**. It changes phase. During melting, the tightly packed atoms or molecules of a solid become moving, rotating, spinning, and colliding liquid molecules. **When substances go from the solid to liquid phase, the atoms and molecules do not change composition.** For example, when ice changes to water, its molecules remain the same composition: two hydrogen atoms to one oxygen atom. What does change is the molecules' amount of kinetic energy. As heat causes ice to melt, the molecules move faster and faster overcoming some of the attractive forces holding them closely together in the solid phase.

The temperature at which a solid melts is its **melting point**. Melting point is a characteristic property of a solid. The melting point of ice is 32°F (0°C). Sugar melts at 365°F (185°C). Salt, on the other hand, melts at 1,474°F (801°C)! Since we rarely, if ever, find conditions that hot, few people have observed salt melting (changing phase from a solid to a liquid).

Teacher Notes: A Solid Start
Lesson Four – Student Reading

Rationale: Hands-on investigations allow students to directly observe chemical phenomena. Readings relate hands-on experiences to vocabulary and new content as well as build literacy skills. Readings also set the stage for future investigations.

Vocabulary Strategy: Word Splash. See Literacy Resource Section for more detail. Word Splash words: atom, molecule, matter, phase, solid, liquid, motion, energy, kinetic, characteristic, property, crystal, lattice, melting point, vibrate, rotate, spin, collide, ice, water

Reading Strategies: Hands-On Investigation, Word Splash. See Literacy Resource Section.
    Before Reading: Lesson Three – A Solid Identification and Word Splash
    During Reading: Word Splash
    After Reading: Word Splash

Materials & Preparation: Student copies of Word Splash & Reading.

Word Splash Teacher's Role: Your role is to monitor the pre-reading and during reading activities and conduct the after reading discussion. In the after reading discussion, highlight the following connections between and among the words.
- **Atoms** and **molecules** are the smallest particles of **matter** that can still be identified as that type of matter.
- **Atoms** and **molecules** have **kinetic energy**.
- **Solids atoms vibrate. Atoms** in **liquids move, spin**, **rotate,** and **collide** with each other.
- When a **solid melts**, its **atoms** gain **kinetic energy** but they do not change form or composition. Note: This is a widely held misconception among students.
- **Characteristic properties** identify **matter.**
- **Ice** is the **solid phase** of **water**.
- **Solids** form **characteristic crystal lattices**. (Some solids are amorphous and do not.)
- **Melting point** is unique to each substance and is a **characteristic property** of **matter.**

Background
In the reading, the terms atom and molecule are formally introduced. The term ion (charged particles) is not introduced because we have found that it confuses middle level students at this point in their learning. A common misconception is that matter is static rather than dynamic. The intention of the reading is to dispel this notion along with another misconception. Middle school students sometimes think that as molecules change phase they change composition or form. A molecule of water is the same form/composition as a molecule of ice or water vapor.

Synergized Middle School Chemistry

Chemistry is confusing to students because interactions between substances can be observed and explained at both the macro- and micro-levels. Students, as well as teachers, frequently have difficulty reconciling what is observed with what is happening at the molecular level. While investigations focus on observations of matter, most of the readings are designed to describe what's happening at the atomic or molecular level during phase changes.

Lesson Five: Melting and Freezing
Teacher Demonstration

Rationale: Providing visual representations is essential for student understanding of molecular motion. If you have teaching aids that animate molecular motion and phase changes, they can be substituted and/or used with the simple demonstration described below.

Materials and Preparation: Styrofoam or other small, lightweight balls of different colors, 8 oz. clear, plastic punch cup

How to Perform the Demonstration

1. Tell the students that atoms and molecules are too small to be seen, even with a microscope. Tell them that using a model can help them visualize how atoms and molecules behave.
2. Place the foam balls in the cup. Tell the students that the foam balls represent atoms. Gently move the cup but do not change the position of the balls to illustrate the vibration of molecules in the solid phase.
3. Ask students to raise their hand if they know what phase of matter is being represented. Call on a student with a raised hand. The anticipated response is: The solid phase because the atoms are not changing position.
4. Now, shake the cup and keep the balls moving. Make sure the balls move around and change location as noted by changes in the position of the colored balls. If a foam ball should escape the cup, that's okay. This represents the process of evaporation.
5. Ask students to raise their hand if they know what phase of matter is represented. Call on a student with a raised hand. The anticipated response is: The liquid phase because the atoms are moving and changing positions but still in contact with each other.
6. Now ask the students what shaking the cup represents. They should connect the shaking to adding energy. Explain that the energy added when a solid changes to a liquid is heat energy. And, that the heat energy is transformed into kinetic energy.
7. Lastly, ask them the term for the phase change from a solid to a liquid. Anticipated response: Melting.
8. Now reverse the process. Instead of shaking the cup to keep the foam balls moving around each other, just shake the cup enough to have the balls move but not change position. Ask the students what phase change just occurred. Anticipated response: Freezing.
9. Ask the students how the atoms changed as the substance went from solid to liquid to solid. They should note that the only change was in their motion. Atoms and molecules do not change form (composition) when they change phase. During a phase change, atoms and molecules change their positions and type of motion.
10. Introduce the idea that even though molecules in the liquid have kinetic energy and are moving, they are not all moving at the same rate. Some move slower than others. Some move fast enough to escape the liquid and move into the vapor phase and evaporation occurs.
11. **Computer Simulation:** If you have access to a computer, have the students observe a simulation of ice melting from a molecular perspective. At the time of this writing, The University

of Illinois' Department of Chemistry has a simulation appropriate for middle school students. Visit http://iclcs.illinois.edu/index.php/chemistry-simulations.

12. Have the students make the following notebook entry: Using pictures, show what happens to atoms and molecules as they change phases from a solid to a liquid and back to solid. Label the drawings. Note: If students are having difficulty starting, suggest that atoms and/or molecules can be represented as circles within a container.

Teaching Tips

1. You can introduce evaporation and condensation by allowing some of the balls to escape the cup and then by putting them back into the cup. Evaporation and condensation are formally addressed in later lessons.

2. Calling on a student with a raised hand increases the probability that you will get a correct response. When teaching new concepts, it is best to reinforce correct responses as much as possible.

Lesson Six: Comparing Water's Properties to Other Liquids
Student Investigation

The Problem: Liquids, like solids, have characteristic properties. In this investigation you will compare and contrast the properties of water with three other common liquids.

Focus Questions
1. What are characteristic properties of water?
2. How does water compare to other liquids?

Pre-lab Preparation: Before you begin testing the liquids, get the general materials listed below and follow the procedure for labeling the liquids.

General Materials (2-3 students per group)**:** Safety goggles, paper towels, masking tape, labeling pen, hand held magnifier, 4 plastic, punch cups: 1 each for water, mineral oil, rubbing alcohol, and vinegar. **Safety Note**: Rubbing alcohol should be used in a well-ventilated area. It will evaporate easily and has a distinct odor.

Labeling Procedure
1. Using masking tape, label each punch cup with the name of the liquid to be tested: water, mineral oil, rubbing alcohol, and vinegar.
2. Go to the supply table and fill each punch cup 1/3rd full of the correct liquid. You will use these liquids in the procedures that follow. If you need more, refill the cup.

Comparing Water's Properties to Other Liquids
Part 1. Water Drops on Different Surfaces
This test uses your observational skills to describe how each liquid looks and moves on two different surfaces: wax paper and plastic wrap.

Focus Question: How does the surface affect the way a drop of liquid looks and moves?

Materials: Safety goggles, general materials, wax paper, plastic wrap, eyedroppers, toothpicks, metric ruler

Procedure 1. Water Drops on Different Surfaces
1. Copy the Observing Liquids on Different Surfaces chart in your science notebook.
2. Wax paper: Use the masking tape to make a label for each of the four liquids. Tape the labels across the top of the wax paper.
3. Using an eyedropper, place a drop of each liquid near its label.
4. Describe the appearance of the drops in your notebook.
5. Using a toothpick, try to drag each drop about 5 cm across the paper.
6. Next, tilt the wax paper so that the drops slide about 10 cm down the paper.

7. Lay the paper flat and wipe off the wax paper with a paper towel. Observe the surface of the wax paper.
8. Repeat procedures 2 through 7 with clear, plastic wrap.
9. Record the observations in your chart.

Observing Liquids on Different Surfaces				
Observations	Water	Mineral Oil	Rubbing Alcohol	Vinegar
Wax Paper				
Plastic Wrap				
Other Surfaces				

Claims and Evidence
1. Which liquid behaves most like water? What is your evidence?
2. Which liquid behaves least like water? What is your evidence?

New Understandings
1. If you had an unknown clear, colorless, and odorless liquid that looked like water, would the surface test be a good way to identify the liquid as water? Why or why not?
2. Which liquid appears the least attracted to wax paper? What is your evidence?
3. Why are cars sometimes waxed after they are washed?

Comparing Water's Properties to Other Liquids
Part 2. Paper Absorption

This test compares how well each liquid is absorbed by two different papers: brown paper (like the paper used to sack groceries) and a paper towel.

Focus Question: How does the type of paper affect the absorption of liquids?

Materials: Safety goggles, general materials, cotton swabs, metric ruler, 4 brown paper squares (5 cm x 5 cm), 4 pieces of paper towel (5 cm x 5 cm)

Procedure 2. Paper Absorption
1. Copy the Absorption of Liquids Chart into your science notebook.
2. Brown Paper: Use a pencil to write the names of the four liquids on each piece of brown paper.
3. Place a cotton swab in each of the four liquids.
4. Rest the swab in the middle of the brown paper square for 30 seconds and remove.

5. Observe how well the paper absorbed the liquid by measuring and recording the diameter of the spot.
6. Try to evaporate the liquid from the brown paper by waving the paper in the air. Record any time differences in evaporation.
7. Repeat the procedures above using paper towel squares.
8. Record the observations in your chart.

Absorption of Liquids				
Brown Paper				
Observations	Water	Mineral Oil	Rubbing Alcohol	Vinegar
Absorption				
Diameter of Spot (cm)				
Evaporation				
Paper Towel				
Observations	Water	Mineral Oil	Rubbing Alcohol	Vinegar
Absorption				
Diameter of Spot (cm)				
Evaporation				

Claims and Evidence
1. Which liquid behaved most like water? What is your evidence?
2. Which liquid behaved least like water? What is your evidence?

New Understandings
1. If you had an unknown clear, colorless, and odorless liquid that looked like water, would the absorption test be a good way to identify the liquid as water? Why or Why not?
2. Which liquid is the most attracted to paper towels? How do you know?
3. Why are some paper towels better at cleaning up spills?

Comparing Water's Properties to Other Liquids
Part 3. Ability to Dissolve Sugar

You found out in the solids investigation that different solids dissolve differently in water. In this test, you will compare how well the same solid (table sugar) dissolves in each of the four liquids.

Focus Question: How does the type of liquid affect how well sugar dissolves?

Materials: Goggles, general materials, table sugar in punch cup, plastic spoons, 4 empty, clean punch cups

Procedure 3. Ability to Dissolve Sugar
1. Copy the chart below in your science notebook.
2. Pour 25 mL of each liquid in a clean punch cup.
3. Add a level spoonful of sugar to each cup. Stir & record your observations.
4. Record the observations in your chart.

Observations of Sugar Dissolving in Different Liquids				
	Water	Mineral Oil	Rubbing Alcohol	Vinegar
Sugar Dissolving Ability				

Claims and Evidence
1. Which liquid behaved most like water? What is your evidence?
2. Which liquid behaved least like water? What is your evidence?

New Understandings
1. If you had an unknown clear, colorless, and odorless liquid that looked like water, would observing the way sugar dissolves be a good way to identify a liquid as water? Why or why not?
2. How could you get the sugar to dissolve faster in water?

Comparing Water's Properties to Other Liquids
Part 4. Mixing Liquids

The way different liquids mix together is a characteristic physical property of a liquid. In this activity you will mix small amounts of each liquid with water.

Focus Question: How well does water mix with other liquids?

Materials: Safety goggles, general materials, food coloring (blue, red, and green), toothpicks, eyedroppers, plain paper to draw a mixing chart, metric ruler, blue, red, and green crayons or colored pencils

Procedure 4. Mixing Liquids
1. Copy the Mixing Water with Other Liquids chart on a plain sheet of white paper.
2. Place plastic wrap over the chart you made in step 1. Hold the plastic wrap in place with masking tape, if needed.
3. Add a few drops of blue food coloring to the water in your cup to identify it as water.
4. Add a few drops of the red food coloring to the rubbing alcohol in your cup to identify it as alcohol.
5. Add a few drops of green food coloring to the vinegar in your cup to identify it as vinegar.
6. Do not add food coloring to the oil.
7. Each square will have two drops of liquid that are not touching. One drop will be water and the other drop will be one of the four liquids.
8. Using a toothpick, drag the water drop in each square to the other drop. As soon as the two drops meet, lift the toothpick without stirring the drops. Observe the way the liquids mix when they contact each other.
9. Record your observations.

Observations: Use blue, red, and green crayons to illustrate how the water mixed with alcohol, vinegar, and oil.

Mixing Water with Other Liquids							
	Water (Blue)		**Mineral Oil (Clear)**		**Rubbing Alcohol (Red)**		**Vinegar (Green)**
Water (Blue)	◯ Water Drop	◯ Water Drop	◯ Water Drop	◯ Mineral Oil Drop	◯ Water Drop	◯ Rubbing Alcohol Drop	◯ Water Drop	◯ Vinegar Drop

Claims and Evidence
1. Which liquid(s) behaved most like water? Use evidence to support your answer.
2. Which liquid(s) behaved least like water? Use evidence to support your answer.
3. Compare your claims and evidence from all four parts of this investigation to at least one other lab group. How do your claims compare and contrast with the other group?

New Understandings
1. What did you learn about liquids by performing the tests?
2. You find a jar with an unlabeled, clear, colorless, and odorless liquid. It looks like water. What tests could you do to support the claim that it is or is not water?
3. After running your tests, could you be sure that the chemical is water? Why or why not?
4. If you repeated the liquid tests, what might you do differently?

Reflections
1. How well did your team work together in this experience?
2. What other comments do you have at this time about what you are learning?

Teacher Notes: Comparing Water's Properties to Other Liquids
Lesson Six – Student Investigation

Rationale: This investigation builds on the students' prior knowledge of liquids as well as provides all students with a common set of observational experiences from which to construct new understandings about the liquid phase of matter. The investigation serves as the before reading activity for the next lesson.

Materials & Preparation: Liquids: water, mineral oil, rubbing alcohol, and vinegar; hand held magnifier, paper towels, wax paper, plastic wrap, eyedroppers, toothpicks, metric ruler, brown paper squares (5 cm x 5 cm), paper towel squares (5 cm x 5 cm), cotton swabs, table sugar, plastic spoons, punch cups, plastic wrap, masking tape, food coloring (blue, red, and green), plain white sheet of paper, blue, red, and green crayons or colored pencils, liquid soap for a clean up.

Keep liquids in their store packaging in a supply area. Oil is messy to use, but its properties contrast well with water. Make sure that you have soap and water ready to clean oily materials.

Teacher's Role: Your role during the investigation is to facilitate lab work by managing the use of supplies, observing student work, and asking probing questions. As you monitor student work, continually assess student understanding as evidenced by their comments, questions, and notebook entries.

Background Information: The focus of the investigation is on the observable differences between water and three other common liquids. A formal discussion of why water (or another liquid) behaves the way it does should be left until after students have completed Lesson Seven - Phasing Into Liquids. The lesson seven reading provides additional background information for this investigation.

About the Liquids: Vinegar is a solution of acetic acid (CH_3COOH) and water (H_2O). Because it is an aqueous solution with about 96% water, it behaves more like water than the other liquids. Rubbing alcohol is also known as isopropyl alcohol (C_3H_7OH). It also is an aqueous solution and behaves similar to water in some circumstances. But, it will not dissolve sugar. Water, vinegar, and rubbing alcohol are all polar compounds. Mineral oil (often called baby oil) is an organic hydrocarbon consisting of 15 to 20 carbons or $(CH_2)_n$ where n is a variable. It is non-polar.

The explanation for water's ability to form "tight" beads is related to the strength of its molecular attractive forces or surface tension.

Key Observations
Part 1: Water Drops on Different Surfaces
Water spreads out on plastic wrap and beads on wax paper.
Neither rubbing alcohol, oil, nor vinegar form as "tight" a bead as water.
Oil does not spread out as much on plastic wrap.

Key Observations
Part 2: Paper Absorption
Water, rubbing alcohol, and vinegar are quickly absorbed by the paper towel. Oil absorbs into the towel very slowly.
Oil is absorbed by the brown paper and forms a translucent spot.
Water, alcohol, and vinegar wet the brown paper and then begin to evaporate. Oil shows no evidence of evaporation.

Key Observations
Part 3: Dissolving Sugar
Sugar dissolves in water and vinegar.
Sugar does not appear to dissolve in rubbing alcohol or oil.

Key Observations
Part 4: Mixing Liquids
Water mixes quickly with water.
Water mixes with both alcohol and vinegar.
Water and vinegar mix slower than water and alcohol.
Water and oil do not mix.

Claims and Evidence Based on Tests
1. **Which liquid(s) behaved the most like water?** Vinegar behaves the most like water. It beads on wax paper, behaves the same on brown paper and paper towels, dissolves sugar, and mixes with water. Rubbing alcohol does not dissolve sugar but behaves similarly in other tests.
2. **Which liquid(s) behaved the least like water? State your claim and use evidence from the lab tests to support your answer.** Oil behaves very differently from water. It doesn't bead on wax paper, it absorbs into the brown paper forming a translucent spot, sugar doesn't appear to dissolve, and oil doesn't mix with water.

Teaching Tips
1. Let students put observations in their own words. Focus on descriptive vocabulary.
2. Encourage students to use drawings to describe the behavior of the liquids. Make sure they appropriately label their drawings.
3. Allow students to try additional procedures, if time and materials permit.
4. This is an excellent activity to have them write a Compare and Contrast Summary.

Lesson Seven: Phasing Into Liquids
Student Reading

(1) Think about what you observed when you compared water to other liquids. Did you notice that vinegar and alcohol behaved like water? Did you notice that oil behaved differently than water? The differences and similarities you observed can be explained by knowing more about the properties of liquids.

(2) **Liquids differ in how their drops cling to different surfaces, how they dissolve solids, their ability to flow (viscosity), and in the way they mix with other liquids.** Because each liquid is composed of different kinds of atoms and molecules, they behave differently. For example, water easily runs off wax paper and mineral oil does not. Also, at room temperature water flows freely, but mineral oil flows very slowly. Thus, mineral oil is more viscous than water. Water easily dissolves sugar, but alcohol does not dissolve sugar. Water mixes with both vinegar and alcohol but not with mineral oil.

(3) **Just like in their solid phase, liquid phase atoms and molecules are attracted to each other by strong forces.** As the temperature of a solid increases, the atoms gain kinetic energy. As energy is added, the atoms overcome some of the force holding them together, and the solid phase changes to a liquid phase. When this happens, the solid melts into a liquid. If you could see the atoms or molecules of a liquid, you would observe them moving around each other, rotating, and spinning.

(4) **The combination of strong attractive forces between atoms or molecules and increased kinetic energy allow liquid phase atoms or molecules to move about, rotate, and spin.** Thus, liquids change shape to conform to their container. When a liquid is poured, the attractive forces keep the atoms or molecules together making it possible to transfer liquids from one container to another. When a liquid spills or is dropped onto a surface, it forms drops on the surfaces or in the air.

Teacher Notes: Phasing Into Liquids
Lesson Seven: Student Reading

Rationale: This reading builds on the previous investigation by providing a basic explanation of the behavior of atoms and molecules during the liquid phase. Students should move from first-hand observations to explaining their observations in terms of atoms and molecules.

Vocabulary Strategy: QAR – Question-Answer Relationships, Everybody Read To…(ERT). See Literacy Resource Section.

Key Vocabulary: Atom, molecule, matter, phase, solid, liquid, motion, energy, kinetic, viscosity, vibrate, rotate, spin, collide, ice, water

Reading Strategies
    Before Reading: Lesson Six – Comparing Water's Properties to Other Liquids
    During Reading: QAR and ERT
    After Reading: Lesson Eight – Student Investigation

QAR Materials & Preparation: Student copies of reading: Phasing Into Liquids

Teacher's Role & Background: QAR and ERT
Paragraph One: Introduce the article by asking your students to read the title and the first paragraph.
1. Tell students to raise their heads and look at you when they are finished reading the first paragraph. Wait until students are facing forward.
2. Prompt students to raise their hands if they can tell the purpose of the reading.
3. Wait until at least 1/3rd of the class raises their hands and then call on a student with a raised hand.
4. Ask a student to read the part of the paragraph that tells the **purpose** of the reading. An accurate responses should state that the purpose is to learn more about the properties of liquids. **From the Reading:** The differences and similarities you observed can be explained by knowing more about the properties of liquids.
5. Before reading further, initiate a discussion to determine what students already know about the properties of liquids by using the questions below as a guide.
    a. In what situations did alcohol behave like water?
    b. How are the properties of water different from oil?
    c. How are liquids different from solids?
    d. What liquids that we did not test might be similar to water? Different from water?
6. Repeat the above procedure with each succeeding paragraph. You may create your own questions or use the QAR questions provided for each paragraph.

Paragraph Two QARs (Note: Some QAR questions can be answered directly from the reading, some require students to make inferences, and some require students to use their background and experiences.)

1. **Describe one property of the liquids we tested and give an example? From the reading:** To cling to different surfaces – water clings better to plastic than wax, oil clings better to wax surfaces; Ability to dissolve or not dissolve solids – water dissolves sugar and alcohol does not; Ability to flow at different speeds – water, vinegar, and alcohol flow freely and oil flows slowly; Ability to mix or not mix with other liquids – water mixes with well with vinegar but neither water nor vinegar mix well with oil.

2. **In your own words, define viscous.** Students should note that viscous in chemistry refers to the flow of liquids. Liquids can be very viscous (thick and gooey) or less viscous (thin and runny) or something in between.

3. **How does the motion of atoms and molecules differ in solids and liquids?** Atoms in solids do not rotate or spin or move around each other like they do in liquids. Atoms in solids vibrate.

4. **Is toothpaste a liquid or a solid? Explain your answer.** Generally speaking, toothpaste is a thick liquid that flows very slowly due to its viscosity. However, it does have some solid properties because it usually contains solid substances that are used as abrasives making it even more viscous. The point of the discussion is to get students to think about how the atoms and molecules are behaving.

Paragraph Three QARs

1. **What holds liquid particles together? From the reading:** Forces within each atomic particle hold the particles together.

2. **At room temperature, ice melts to form water. Give examples of substances that are solid at room temperature and change to a liquid if left in the hot sun.** Students usually think of food items like chocolate, butter, and hard sugar candies. They may also mention some plastics and tars. Emphasize that atoms and molecules do not change as solids melt, only their motion.

Paragraphs Four QARs

1. **Explain why liquids take the shape of their container and can be poured? From the reading:** The combination of strong attractive forces between the particles and increased kinetic energy allow atomic particles in the liquid phase to move about. …When a liquid is poured, the attractive forces keep the substance together **…**

2. **Is it possible to break the attractive forces that hold solid and liquid atoms and molecules together? How could it happen?** If enough energy is added to the liquid by heating or other means, the forces can be broken and the liquid atoms and molecules escape into the air or gas phase. This might be a good time to mention that not all atoms and molecules are moving at the same speed. Some move faster than others.

Teaching Tips

1. Synergetic lessons have a greater probability of increasing student achievement if they are taught in the prescribed sequence. In this case, preceding this reading with lesson six is critical to learning because lesson six is both a hands-on investigation and a before reading strategy.

2. The vocabulary from the word splash is now presented in the student reading to reinforce key vocabulary.

3. The ERT strategy can be differentiated to meet the needs of your class. Students can be partnered with another student or helping adult if they cannot read by themselves. However, if students only read with partners, they do not get the practice they need to become fluent nonfiction readers on their own. Use your judgment.

4. **Computer Simulation:** If you have access to a computer, have the students watch a simulated separation of oil/water mixture and a salt crystal in water from a molecular perspective. At the time of this writing, The University of Illinois' Department of Chemistry has a simulation appropriate for middle school students.

Visit http://iclcs.illinois.edu/index.php/chemistry-simulations.

Lesson Eight: Observing Kinetic Energy of Liquids
Student Investigation

Although you can't see atoms and molecules, you can indirectly observe the results of their collisions.

Focus Question: How does food coloring behave in different water temperatures: cold, room temperature, and warm?

Prediction:

Materials: Safety glasses, 3 punch cups, red food coloring, eyedropper, cold water, room temperature water, and warm tap water. CAUTION: The water should be warm, not hot.

Procedure
1. Fill each punch cup ¾ full of warm tap water, room temperature water, and cold water.
2. In your science notebook, draw three cups like the ones shown below. You will use them to record your observations.
3. Quickly add one or two drops of red food coloring to each cup.
4. After 30 seconds, draw your observations in your science notebook.
5. Continue to observe for 2 minutes.
6. After 2 minutes, describe how well the food coloring has mixed with the water.

Warm Water   **Room Temp**   **Cold Water**

Claims and Evidence
1. How did your prediction compare to your observations?
2. What evidence supports the claim that temperature affects the behavior of atoms and molecules?

New Understandings
1. Diffusion is the mixing of one kind of molecule with another. In this case, food coloring diffused into the water. Write a statement relating the rate of diffusion in a liquid to the temperature of the liquid.
2. Describe what you think each of the cups would look like after one hour.

Reflections: Other questions or comments.

Teacher Notes: Observing Kinetic Energy of Liquids
Lesson Eight: Student Investigation

Rationale: Relating the movement of atomic particles to the behavior of matter's phases requires students to reconcile what they can see to what they can't see. This activity uses indirect evidence to support the claim made in the reading that liquid particles are constantly moving. Students will compare the mixing of food coloring with warm water, room temperature water, and cool tap water.

Materials & Preparation: 3 punch cups, red food coloring and eyedropper, cold water, room temperature water, and warm tap water. Note: If food coloring is available in the squeeze tip bottles, you don't need eyedroppers. **Safety Caution:** It is not necessary to have the water hot. In fact, if the water is too hot it could burn the students and even begin to melt the cup. If ice is available, you could make a pitcher of ice water; however, don't have ice cubes in the water when the food coloring is added. Use your judgment.

Teacher's Role: Your role during the investigation is to facilitate lab work by managing the use of supplies, observing student work, and asking probing questions. As you monitor student work, continually assess student understanding as evidenced by their comments, questions, and notebook entries.

Background: Students will see a definite difference in the mixing speed of food coloring with water at each temperature. The warmer the water, the faster the food color will mix. Depending on class time, the food color may not mix thoroughly in the cold water. As an extension or homework assignment, students can be asked to place a cup of cold water and food coloring in a refrigerator to see if it will mix thoroughly if given enough time. It will mix thoroughly.

New Understandings
1. **Diffusion is the mixing of one kind of molecule with another. In this case, food coloring diffused into the water. Write a statement relating the rate of diffusion of a liquid to the temperature of the liquid. Possible response:** A liquid like food coloring will diffuse faster into another liquid if it is warmer. The warmer the liquid, the faster the rate of diffusion. Note: Diffusion will be studied again in Lesson Fourteen.
2. **Describe what you think each of the cups would look like after one hour. Possible correct response:** In time, all of the cups will look the same and be the same color shade, depending on the food coloring. Eventually, all the molecules will mix uniformly. Note: Once the food coloring molecules have thoroughly diffused into the water, the water molecules and food coloring molecules are still moving about, spinning, and rotating.

Teaching Tips
1. This investigation can also be done as a demonstration. Using large, clear containers can be very dramatic.

2. Make sure at least one set-up remains until the food coloring fully diffuses into each liquid. This could be by the end of the class or overnight.

Lesson Nine: Explore Some More – Evaporation Inquiry
Student Investigation

Introduction: Atoms and molecules of the same substance gain different amounts of kinetic energy when heated. If you could see a liquid's atoms or molecules, you would see some of them moving faster than others. Near the surface, some atoms or molecules would be escaping into the air.

The natural process of atoms or molecules at the liquid's surface escaping into the air is called **evaporation**. For example, if a dish of water stands overnight, there will usually be less water in the glass in the morning. Why? More water molecules near the surface have escaped into the air than water molecules from the air entered the liquid. Water molecules that escape are now water vapor molecules because they are in the vapor phase. Oil molecules require much more energy to escape into the air; thus, the evaporation of oil takes place much more slowly than water.

Focus Question: What variables increase or decrease the rate of evaporation of liquids?
Note: Your experiment is limited to the time frame determined by your teacher.

Your Focus Question:

Your Prediction:
Possible Materials per Group: Goggles, masking tape, measuring tools: spoons, balance, thermometers, beakers, clock or timer, and/or graduated cylinders, containers to hold liquids
Water (warm to cool) and other liquids like rubbing alcohol, vinegar, oil
Solutions of salt water, sugar water, borax, and/or Epsom salt
Other materials: as requested by students and available for use.

Procedure: In your notebook, write the procedure you will use to determine the effect of your variable on the rate of evaporation of water. Have your procedure checked by your teacher before you begin your set-up.

Observations/Data: Record your observations and data in your science notebook. Display your results in appropriate form: chart, graph, pictures and/or words that can be shared with the class.

Claims and Evidence: State your claims based on your evidence and prepare to present your findings to the class.

New Understandings: Complete after class presentations.
1. What did you learn about evaporation?
2. What other questions do you have about evaporation?
3. What did you learn about conducting a scientific investigation?

Reflections: Evaluate your work in comparison to others in the class?

Teacher Notes: Explore Some More – Evaporation Inquiry

Lesson Nine: Student Investigation

Rationale: Up to this point, the investigation lessons have guided student lab work and were designed to familiarize students with the basic parts of an investigation. Now, students will apply their lab skills to design their own evaporation inquiry.

Materials Preparation: Various types of containers, tap water (warm to cool) and other liquids like rubbing alcohol, vinegar, oil, masking tape; pencils for labeling; various solids like table salt, sugar, borax, and Epsom salt; measuring tools: spoons, balance, thermometers, beakers, clock or timer, graduated cylinders or beakers.

The amount and type of materials available to students will depend on each classroom setting. Before students begin their inquiry design, let them know what you have available. The basic materials are those used in the preceding labs. Solid substances are included for students to add to a liquid, if they desire. Make sure you approve all student lab designs before they begin. Make sure procedures are safe and students know all appropriate precautions.

Teacher's Role: After students read the introduction, brainstorm variables that might affect evaporation. As students design their inquiry, they will need to think about the variables that may increase or decrease the kinetic energy of the molecules at the surface. Variables to consider are: temperature of the liquid, temperature of surroundings, air currents, surface area for evaporation, type of liquid, movement of liquid, and type of container.

Before beginning, introduce the idea of a control – something to compare to their test. Give an example of a control from Lesson Six – Comparing Water's Properties to Other Liquids. For example, when we mixed blue water drops with the drops of three other liquids, we also mixed blue water drops with blue water drops as a basis for comparison.

Have the students work in groups. This allows for differentiation as well as reduces supply needs. Stress the need to select one variable and design a test to show how it affects evaporation.

Claims and Evidence Presentation: As students present, listen for the way they use scientific terms and evidence to support their claims. Guide students to a deeper understanding of evaporation by asking students to explain their findings in terms of what the molecules are doing. For example, if a group claims that warmer liquids evaporate faster than cooler liquids, help them understand that the molecules at the surface of a warm liquid have more energy than the same molecules on the surface of a cool liquid; thus, the warmer the liquid, the faster the rate of evaporation.

Background: Evaporation is phase change that is frequently misunderstood by both students and adults. Evaporation occurs at the surface of a liquid. Because the atoms and molecules of a liquid do not all have the same amount of kinetic energy, some atoms and molecules at the surface have enough energy to escape the liquid. Likewise, some of the gas phase atoms and molecules may lose energy and return to the liquid phase (condensation). The focus of instruction should be the behavior of the molecules at the surface. When molecules escape from the liquid to the gas phase faster than the gas phase molecules enter the liquid, the rate of evaporation increases. Variables likely to change the rate of evaporation are: the temperature of the liquid, the surface area of the liquid, whether or not the liquid is stirred, the kind of liquid, and whether or not the liquid's container is covered.

If a group chooses to set up an investigation to test the rate of evaporation in a container with a lid vs. no lid, you will need to introduce them to the concept of a closed system. In a closed system, evaporation and condensation will reach equilibrium. Equilibrium is reached when the number of molecules leaving the liquid (evaporating) equals the number of molecules entering the liquid (condensing).

Teaching Tips

1. This is a good investigation to focus on the concepts of independent variable, dependent variable, and constants. A variable is a factor that takes on different values in an investigation. Variables can be purposely changed (independent) while other variables (dependent) respond to the change. For example, if the temperature of a liquid is changed, then the rate of evaporation responds to the change. Students should have an experimental set-up and a control. Both set-ups should differ by only one variable. Constants are those factors that remain the same for both set-ups. For example, if temperature is the independent variable and amount of liquid remaining after a given time the dependent variable, some of the constants would be: type of containers, kind and amount of liquid, and location of the containers.

2. A common misconception is that molecules change form when they change phase. Stress the point that liquid molecules are the same as their gas or vapor molecules. The only difference is the speed that the molecules are moving.

3. To differentiate learning, you can vary the size of the groups, place students together that have complimentary skills, and design groups so that you can monitor and assist students with special needs.

4. **Assessment Opportunity:** This investigation makes an excellent performance-based assessment. If you are going to use this as a performance-based assessment, make sure that the students know the evaluation criteria before starting the lab. Whether using this as a formal assessment or not, it is always a good idea to have students self-evaluate investigations of their own design.

Lesson Ten: Condensation
Student Investigation

At the same time atoms and molecules are evaporating, some are condensing. **Condensation** refers to the process of gas molecules returning to the liquid phase.

Focus Question: How does temperature affect condensation?

Prediction:

Materials (per group): 3 clear plastic cups, plastic wrap, 2 rubber bands, very warm tap water, 1 ice cube, and paper towels

Procedure:
1. Fill all three cups ½ full of very warm tap water.
2. Cover two cups with plastic wrap and secure the plastic wrap with a rubber band.
3. On the top of one of the plastic covered cups, place an ice cube.

Plastic Wrap & Ice Cube   Plastic Wrap Only   Very Warm Water
Very Warm Water   Very Warm Water

Observations: Carefully, observe the inside surface of the cups and the plastic wrap. In your notebook, draw a picture comparing the three cups. Carefully, note the size of any drops that form on the cup and/or the plastic wrap.

Claims and Evidence: Make a statement relating condensation to temperature change and support it with evidence.

New Understandings
1. When you have a cold drink, sometimes water forms on the outside of the bottle or glass and people say, "The glass is sweating!" Do glasses really sweat? Explain.
2. When you take a shower, sometimes the bathroom mirror fogs up. Why?
3. You can't see water vapor, but you can see liquid water. What are clouds?

Reflections: What other questions do you have about either evaporation or condensation?

Teacher Notes: Condensation
Lesson Ten: Student Investigation

Rationale: This investigation provides observational evidence for the process of condensation. It reinforces previously introduced concepts and sets the stage for the next investigations on the properties of gases.

Materials & Preparation: Punch cups, rubber bands, ice cubes, very warm tap water, paper towels. Note: The water does not have to be very hot for this to work well.

Teacher's Role: Your role during the investigation is to facilitate lab work by managing the use of supplies, observing student work, and asking probing questions. As you monitor student work, continually assess student understanding as evidenced by their questions, and notebook entries.

Background: Heating water increases the rate of evaporation. When plastic wrap is placed over the top of the cup, the water vapor is now trapped in the top ½ of the cup. When the ice is placed on top of the plastic wrap, the top ½ of the cup is cooled. The water vapor molecules lose kinetic energy and condense on the cooler surfaces inside the cup. The drops are largest on the plastic wrap closest to the ice. In fact, the drops may get so large that it could rain! What can't be seen are the water vapor molecules that are returning to the water (condensing) at the water's surface.

Observation: When the ice is placed on the top of the plastic wrap, small drops of water appear on the inside top ½ of the cup. Larger water droplets appear on the inside of the plastic wrap just under the ice cubes.

Claims and Evidence
Make a statement relating condensation to temperature change and support it with evidence. As air cools, the rate of condensation of water vapor in the air increases. The cup with the ice cube had the most water condensing inside the cup.

New Understandings
1. **When you have a cold drink, sometimes water forms on the outside of the bottle or glass and people say, "The glass is sweating!" Do glasses really sweat? Explain.** Although people say their drinks "sweat," that isn't the case. It isn't the water from inside moving through the glass, plastic, or aluminum. It is actually the water vapor in the air on the outside of the drink condensing on the cooler outside surface of the container. If the air is very humid (contains a lot of water vapor), much more vapor will condense than if the air is dry.
2. **When you take a shower, sometimes the bathroom mirror fogs up. Why?** As the shower water evaporates, it condenses on the cooler mirror.

3. **You can't see water vapor, but you can see liquid water. What are clouds?** Clouds are millions of water droplets or ice crystals that are suspended in the air. Clouds form when water vapor condenses on dust particles. You can't see water vapor. Note: Steam is not water vapor because you can see it. Steam consists of droplets of liquid water.

Lesson Eleven: Air Matters
Student Investigation

Introduction

In the readings, you learned that solids and liquids are two phases of matter. In this lesson, you will learn more about a third phase of matter – the gas or vapor phase.

Although air is all around us, "air" is usually not the first thing you think of as an everyday chemical. Most students overlook air because the common gases in air are invisible, colorless, and usually odorless; yet, we can't live without air. Air is actually a mixture of gases. It is approximately 78% nitrogen gas, 21% oxygen gas, and 1% other gases like carbon dioxide and water vapor, to name a few.

Today, we take the existence of the gases in the air for granted. But, this wasn't always the case. It wasn't until the second half of the Eighteenth century that Joseph Priestly identified and provided evidence for the existence of oxygen in the air. In this lesson, you will use your previous knowledge and experiences to support the existence of air as matter.

Focus Question: If air is invisible, how can you prove that air exists?

Your Challenge: To provide convincing evidence to support the claim that air exists as matter. Note: Matter is defined as anything that takes up space and has mass.

Materials: Butcher paper, crayons, markers, construction paper, scissors, tape and other miscellaneous items

Brainstorm

On Your Own: In your notebook brainstorm a list of observations that provide evidence for the existence of air as matter.

Pair-Share: Work with a partner to combine lists and add any additional observations.

Design and Present an Air is Matter Poster: Working with your partner, use your list and design an Air is Matter Poster. Be prepared to present your poster to the class. Make sure you can explain how your poster shows that air has both mass and takes up space.

New Understandings. To be completed after poster presentations.
1. What three pieces of evidence best illustrate that air takes up space (volume)?
2. What three pieces of evidence best illustrate that air has weight (mass)?
3. How did your evidence compare to the evidence on other posters?
4. What did you learn about the gases in air that you did not know?

Reflection: Why is it important to share scientific ideas?

Teacher Notes: Air Matters
Lesson Eleven: Student Investigation

Rationale: This lesson uncovers student ideas about gases. Student-to-student discussions and presentations motivate reluctant learners and foster a collaborative learning atmosphere.

Materials & Preparation: Butcher paper, crayons, markers, construction paper, scissors, tape and other miscellaneous items like balloons, feathers, and straws could be added to posters.

Teacher's Role: Your role during the investigation is to facilitate student work by managing the use of supplies, observing student work, and asking probing questions. As you monitor student work, continually assess student understanding as evidenced by their comments and questions. When groups are ready to present, tell the groups that each person has to speak and be part of the presentation. Presentations should be about 5 minutes.

Background: Students frequently don't think of air as a chemical because it is invisible to them. Students come up with many pieces of evidence. Some common examples of evidence are listed below. Students need to relate their evidence to the definition of matter – either showing air has volume or air has mass or both.

Common Poster Examples
- **Wind** shows air has mass – example: wind shows that the force of air can produce energy to move objects like a wind turbine or windmill. Note: force equals mass times acceleration
- **Air's resistance** to moving objects shows air has mass – example: race cars are designed to reduce air resistance or the force of air on the car
- **Air supports objects** (mass and volume) – example: air planes
- **Balloons** filled with different gases behave differently showing that air has both mass and volume – example: helium balloon vs. air-filled balloon
- **Bubbles** show that air has both volume and mass – example: carbonated beverages weigh less after fizzing; air expands lungs (volume); used in tires (volume); some gases are visible and have an odor (mass) – example: air pollutants like sulfur dioxide
- **Air can be measured** (volume & mass) – example: pressure gauges for tires.

Teaching Tips
1. **Teacher Demonstration:** Begin or end this lesson with a teacher demonstration. A popular demonstration uses a piece of waded tissue, secured to the inside bottom of a drinking glass. The glass is turned upside down and immersed in a fish tank or other tub of water. Many students think that the water will rise in the glass and wet the paper. Air keeps that from happening.

2. This lesson is best done spanning two class periods. One class period is used to design the poster and the second for presentations. Students can also design their posters outside of class.

3. Although students can easily memorize the definition of matter, they have difficulty applying it to gases. Be prepared to help students understand how their examples show that gases have mass and/or volume.

4. **Extension:** Select one example from each poster for the students to demonstrate to the class. For example, for air resistance, have student's show that a piece of waded paper falls faster than a flat piece of paper due to air resistance. Follow up by asking students to predict if the papers will behave the same in a vacuum? You can let them discuss and wait to answer this question after the next lesson.

Lesson Twelve: Gases and Temperature Change
Student Investigation

Purpose: To observe what happens when air changes temperature.

Focus Question: How does temperature change affect the volume of air?

Materials

> Empty plastic bottle like the ones sold with water or pop
> 3-plastic, punch cups
> Liquid dish detergent
> Very warm tap water
> Cold tap water

Procedure
1. Pour just enough liquid dish detergent into a punch cup to cover the bottom of the cup.
2. Fill a second punch cup ½ full with very warm, tap water.
3. Fill a third punch cup ½ full with cold, tap water.
4. Remove the lid from the plastic bottle.
5. Dip the opening of the bottle into the liquid dish soap to form a film over the mouth of the bottle.
6. Place the bottom of the bottle in the hot water cup. Observe.
7. Remove the bottle from the hot water and place it into the cold water. Observe.
8. Repeat at least three times. If the detergent film breaks, dip the bottle again.

Observations: Describe what happens to the soapy film as the bottle moves from hot water to cold water and back again.

Claims and Evidence: Make a statement relating the temperature inside the bottle to the volume of gas. Support your statement with evidence.

New Understandings
1. What happens to a balloon that is filled with air and then placed in a refrigerator?
2. Why is it a good idea to leave the window of a car slightly open when it is parked in the sun?
3. Can you think of other examples that show the relationship between temperature and volume of a gas?

Reflections: Did this activity help you learn more about gases? Why or why not?

Teacher Notes: Gases and Temperature Change
Lesson Twelve: Student Investigation

Rationale: This high-interest investigation illustrates the basic relationship between the temperature and volume of gases.

Materials (Per Group) & Preparation: 1-16 oz. or smaller empty plastic bottle (e.g., water or soda bottles), 3-plastic, punch cups: one with a small amount of liquid dish detergent, one with very warm tap water, and one with cold tap water. The bottom of the plastic bottle must be small enough to fit inside the punch cup when it is submerged in the water. The soap solution needs to be thick enough to form a film over the mouth of the empty bottle as it is dipped in the solution.

Teacher's Role: Your role during the investigation is to facilitate lab work by managing the use of supplies, observing student work, and asking probing questions. As you monitor student work, continually assess student understanding as evidenced by their questions, and notebook entries.

Background Information: Charles' Gas Law states that the volume of a gas is directly proportional to the temperature of the gas when pressure is kept constant, and the amount of gas doesn't change. Thus, as temperature is increased, the volume of the gas increases. This is a basic property of gases. Middle school students should not be assessed on their memorization of Charles' Law but on their understanding of the relationship between temperature and volume. This as well as the other gas laws will be reintroduced in high school.

Claims and Evidence: Make a statement relating the temperature inside the bottle to the volume of gas. Support your statement with evidence. As temperature increases, volume increases. As temperature decreases, volume decreases. There is a direct relationship. This law applies when pressure and the amount of gas remains the same. As we heated the air in the bottle, the bubble film expanded over the top of the bottle. As we cooled the air, the film moved down into the bottle. Although the volume changed, the amount of gas remained the same.

New Understandings
1. **What happens to a balloon that is filled and then placed in a refrigerator?** The balloon's volume will decrease and the balloon will get smaller.
2. **Why is it a good idea to leave the window of a car slightly open when it is parked in the sun?** The sun heats up the air molecules inside the car. As they gain kinetic energy, the temperature rises. If a window is open, the air molecules can escape reducing the temperature inside the car. However, in some places, cars can still become dangerously hot if the windows are not open wide. Pets and even humans have died or become seriously ill with heat stroke when left in an unvented car.
3. **Other possible examples:** Soccer balls, footballs, and playground balls inflated in the summer will shrink in the colder months; balloons will increase in volume if filled indoors and then put in the sunlight; pool floats will expand when left in the sun; soda or other carbonated drinks left in the sun can "erupt" when opened.

Lesson Thirteen: Comparing Matter's Phases –
Feature Analysis
Student Reading

Before Reading: Place a √ in the square or squares of the correct phase of matter described by the behavior of their atoms or molecules. Use information learned in previous lessons.

Semantic Feature Analysis Chart Comparison of Matter's Phases with Behavior of Atoms and Molecules									
Features: Behavior of Atoms and/or Molecules									
Phase of Matter	Not moving but have energy and vibrate in place, not changing position	In constant motion	Held together by attractive forces	Spinning, rotating, and moving around each other, changing position	Far apart and moving independent of each other	Arranged in a crystal lattice	Movement increase with temperature	Molecules have the greatest kinetic energy	Exerts pressure on the walls of the container
Solid									
Liquid									
Gas									

Lesson Thirteen: Comparing Matter's Phases
Student Reading

In gases, molecules are far apart and move independently of other atoms or molecules. If you could see gas molecules, they would be shooting around, spinning, and colliding with each other and anything else that gets in their way. In liquids, molecules vibrate, spin, and move around each other. In solids, molecules vibrate, but do not move about.

If you drop a solid object, like a book, on your foot, you feel the pressure of the book's molecules pushing on your foot. Liquid molecules also exert pressure on the surfaces they contact as they spin and whirl. You can feel water's pressure when you are in a swimming pool, especially if you dive beneath the surface. Likewise, as gas molecules collide with surfaces, they exert pressure on surfaces. Right now, billions of gas molecules are bombarding every inch of your skin. Unlike the book and water, you don't usually notice air's pressure because your body has adjusted to the Earth's relatively constant atmospheric pressure.

In the soap film investigation, you observed *the effect of temperature on the volume of a gas (air) when pressure remains constant.* The warm water increased the temperature inside the bottle. The air molecules inside the bottle gained kinetic energy and moved more rapidly. As they bombarded the soap film, the soap film expanded forming a bubble over the top of the bottle. *The volume increased as the temperature increased.* When the bottle was placed in the cool water, the molecules lost kinetic energy, and the film contracted moving below the bottle's opening. *The volume decreased as the temperature decreased.*

In the soap film investigation, both *the number of molecules inside the bottle and the air pressure inside and outside the bottle remained the same.* When we changed the temperature, only volume changed. The air pressure inside the bottle and outside the bottle remained constant as the soap film expanded and contracted.

However, if the bottle had been capped, the pressure inside the bottle would increase (the pressure outside remains the same). A capped bottle cannot expand when heated. If the volume cannot change, the increased temperature will increase the pressure inside the bottle. Can you think of a situation where the pressure inside a bottle is greater than outside the bottle? What will happen when the bottle is opened?   Think soda pop!

Teacher Notes: Comparing Matter's Phases
Lesson Thirteen: Student Reading

Rationale: This reading builds on the previous lessons and provides a basic explanation for the behavior of atoms and molecules during the solid, liquid, and gas phases. As students compare and contrast each phase, they deepen their conceptual understandings.

Strategy: Feature Analysis Chart. See Literacy Resource Section.
    Before Reading: Previous Lessons, Feature Analysis Chart
    During and After Reading: Feature Analysis Chart

Materials: Student page with feature analysis chart; Student reading: Comparing Matter's Phases; Previous readings – Lesson Four: A Solid Start and Lesson Seven: Phasing into Liquids

Teacher's Role
Have students **check (√)** the appropriate boxes based on their previous knowledge. Clarify any chart vocabulary words before the reading. Have students place an **X** in boxes during and after the reading. Make sure students read each paragraph carefully to determine what information is referred to in the chart. It is suggested that first students read on-their-own and then pair-share with their reading buddy. Make sure each student uses a √ for the before reading and an X for during and after reading. Have students circle any features that they want to discuss further. After discussion, collect their charts to assess their current understanding. Return their charts and discuss any additional features before students take the post-assessment. The feature analysis chart will make an excellent study guide.

Background: Below is a completed feature analysis chart for your reference.

Semantic Feature Analysis Chart Comparison of Matter's Phases with Behavior of Atoms and Molecules									
Features: Behavior of Atoms and/or Molecules									
Phase of Matter	Not moving but have energy and vibrate in place, not changing position	In constant motion	Held together by attractive forces	Spinning, rotating, and moving around each other, changing position	Far apart and moving independent of each other	Arranged in a crystal lattice	Movement increase with temperature	Molecules have the greatest kinetic energy	Exerts pressure on the walls of the container
Solid	√		√			√	√		
Liquid		√	√	√			√		√
Gas		√		√	√		√	√	√

Lesson Fourteen: Do You Have a Super Nose?
Student Investigation

During the **process of diffusion** one kind of molecule mixes with another. The molecules that are diffusing move from the place they are highly concentrated to places where they are less concentrated. Eventually, they thoroughly mix with the other molecules. In this lesson you will use your nose to detect scented molecules on the move. The perfume industry employs people known as perfumers, sometimes referred to as "super noses." The "noses" are able to detect many different fragrances with accuracy! It is reported that people with this above average sense of smell are paid very well for their talents.

Focus Questions:
1. Do you have a "super nose"?
2. How well can you identify common fragrances?

Materials per group: Scented balloon (one per station) **Caution:** If you are hypersensitive to smells, inform your teacher for an alternative activity.

Procedure
1. Make a copy of the chart below in your science notebook. Add the same number of rows as the number of balloon stations.
2. Follow your teacher's directions and go to your balloon station.
3. Smell the balloon and predict the scent. Record your prediction. Make sure you record your observations in the correct numbered row. **For example, if you start at station 3, make sure you record your response in row 3.** Keep your prediction a secret and do not reveal your answer to other groups. It's more fun that way!
4. When prompted by your teacher, move to the next station. Record your prediction.
5. After you have been to all the stations, your teacher will help you check your predictions to see if you have a superior sense of smell.
6. Record your observations in your science notebook.

Fragrance Identification		
Station Number	Scent Prediction	Other Comments
1		
2		
3		
4		
5		
6		
7		
8		

Claims and Evidence. Complete after class discussion.
1. How many scents was your group able to identify?
2. What was the most number of scents that any group correctly identified?
3. What is the average number of scents correctly identified?
4. Can you claim that you have a "super nose"? What is your evidence?

New Understandings
1. Was your sense of smell better at the beginning or end? Explain.
2. Do you think you would get better with practice? Explain.
3. Why can you smell the scent placed inside a balloon?
4. Was this a good experiment? How could you improve it?

Reflections and Other Comments: What made it easier to identify some scents rather than others?

Teacher Notes: Do You Have a Super Nose?
Lesson Fourteen: Student Investigation

Rationale: This lesson uses novelty to introduce the concept of diffusion and serves as the before reading strategy for the next lesson.

Materials: Balloons, eyedroppers, extracts, masking tape, and pencil/marker

About the extracts: You will need to gather a variety of common extracts – one scent for every station or student group. The following are readily available in the baking section of most supermarkets: vanilla, chocolate, pineapple, lemon, coconut, almond, cherry, strawberry, mint, and orange. There are more than 57 extracts commonly available.

Station Set-Up: You can either prepare the balloons or have students prepare them.

Teacher-prepared balloons: Extracts evaporate quickly so you need to prepare the balloons close to the start of the class.

1. Add 2-3 drops of an extract to a balloon before you blow it up.
2. Blow up the balloon and attach a piece of masking tape to the tied off end and number the balloon.

Student-prepared balloons: Another method is to let students prepare the balloons.

1. Direct students to their stations.
2. Provide them with a balloon, eyedropper, and extract. Tell them not to reveal the identity of the extract to other groups. In this case, every student will know the identity of at least one extract.
3. Direct the students to add 2-3 drops of extract to a balloon before they blow it up.
4. Show students how to attach a piece of masking tape to the tied off end and number the balloon.

Storing prepared balloons: Because the extracts immediately begin diffusing through the balloon's surface, you will want to store the balloons in a large plastic trash bag that can be sealed until use.

Option: Teacher Demonstration

The balloon investigation is a very high interest activity for students. If you choose not to do the full activity, start the reading lesson that follows with the following demonstration.

1. Using two, extract-filled balloons, ask for student volunteers to identify the scents.
2. Ask the class why you can smell the scents on the outside of the balloon when the scent was placed inside.
3. From here you can transition into the reading.

Teacher's Role: Your role during the investigation is to facilitate lab work by managing the use of supplies, observing student work, and asking probing questions. As you monitor student work, continually assess student understanding as evidenced by their comments, questions, and notebook entries.

Background: Diffusion occurs when molecules move from an area where they are highly concentrated to an area of lower concentration. As molecules diffuse, they mix with other molecules. When the molecules are thoroughly mixed, they are distributed equally throughout the space. And, they are still moving.

Claims and Evidence

1. **What was the most number of scents that any group correctly identified?** Occasionally, a group gets all the scents correct. This is unusual because some of the extracts are not as representative of the scent as they should be.
2. **What is the average number of scents correctly identified?** With 10 scents, most groups should get half of them correct.
3. **Can you claim that you have a "super nose"? What is your evidence?** Probably not. The conditions were not ideal. And, you'd have to identify many more scents. However, students will differ in their abilities and some will tell you that they are much more sensitive to odors and smells than others.

New Understandings

1. **Was your sense of smell better at the beginning or end?** The beginning. The sense of smell "fatigues" as more fragrances are introduced.
2. **Do you think you would get better with practice?** Answers vary.
3. **Why can you smell the scent through the balloon?** Use student ideas to transition to the reading that follows. Let the students know that the balloon allows air molecules and scent molecules to move in and out through openings that are invisible to the eye but large enough for molecules to escape. The reading that follows provides more information.
4. **Was this a "good" experiment? How might you improve it?** Answers will vary; however, students should recognize that we did not have a control (balloon with only air) or a very large sample size. Also, the environmental and testing conditions were variable from group to group.

Teaching Tips

1. If possible, use a long hallway, gymnasium-type space, or an outdoor setting.

2. Some students are very sensitive to smell. Describe what students will do and have a plan for students who choose not to do the station activity. These students can participate in the discussion. The station activity usually takes no more than ten minutes. It's the discussion that follows that is most important.

3. For the discussion phase, begin by asking a student group for their response. Then ask if there are any other responses. If there are, try to come to an agreement on the correct scent. Then, tell the class the correct response.

4. **Prizes Optional:** Students love this investigation. If possible, it is fun to award the groups with the best noses a prize like a crazy pencil or piece of candy.

Lesson Fifteen: Diffusion
Student Reading

Five Most Important Words: As you read, select the five most important words in the paragraphs below. In your notebook, write the words and the reason for your selection.

Reading: You were able to detect the scents from each balloon because the gas molecules escaped from the balloon. The balloon's surface is actually covered with thousands of openings too small for the eye to see. The openings are large enough to let the scent molecules pass through. That's why balloons can deflate even though they aren't punctured.

As the scent molecules passed through the balloon's skin, they diffused into the air. Diffusion refers to the mixing of one kind of atom or molecule with different kinds of atoms or molecules. When molecules diffuse, they move from an area of high concentration to an area of low concentration. For example, the scent molecules escaped the balloon and began mixing with the air molecules. The scent molecules moved away from the balloon's surface where they were highly concentrated. Eventually, the scent molecules and air molecules will be evenly mixed or dispersed.

Here's another example. If you open a bottle of perfume at one end of the room, the perfume molecules will begin randomly mixing with air molecules. Students sitting closest to the perfume bottle will smell the fragrance first. Those sitting farthest from the bottle will smell it last. Eventually, the perfume molecules will be evenly mixed with the air molecules as all the molecules move rapidly about colliding with each other.

In an earlier lesson you observed food coloring added to cold, room temperature, and warm water. This is another example of diffusion. The food coloring molecules mixed with the water molecules until the entire solution was uniformly colored. The food coloring in the cold water took the longest to diffuse. In cold water the molecules move slower than warm. When temperature increases, molecules move faster. The same is true for gas diffusion. The warmer the temperature, the faster gases will diffuse.

If you puncture a balloon, the gases escape rapidly. The rapid expulsion of gases is called effusion. Effusion also happens when a tire is punctured or when rocket fuels ignite. In rockets, gases are quickly released providing the force needed to propel the rocket.

Teacher Notes: Diffusion
Lesson Fifteen: Student Reading

Rationale: In previous readings, important words were highlighted. In this reading, words have purposely not been highlighted. It is now time for students to determine the important words.

Strategies: Five Most Important Words, Three-Part Vocabulary. See Literacy Resource Section.
Before Reading: Lesson Fourteen: Do You Have a Super Nose? Or Demonstration (see Lesson Fourteen Teacher Notes for optional Teacher Demonstration)
During Reading: Five Most Important Words
After Reading: Five Most Important Words, Three-Part Vocabulary

Five Most Important Words Teacher's Role: This strategy is used both during and after reading. Students will find that they must read the text a number of times. Astute readers should select **gas, diffusion**, and **concentration** as three of the five words. If students did not select these words, make sure the words are included during class discussion. Ask the class why these three words are important in the reading. Stress the connection between the reading's purpose and the important words selected. Discuss all other words selected and solicit a variety of student responses. The term effusion is introduced to contrast to the more passive process of diffusion.
After reading: Have the students pair-share their five words with a literacy partner and the class.

Three-Part Vocabulary Teacher's Role: Have the students define diffusion in their science notebooks by listing the word, a definition in their own words, and a labeled picture providing an example of the concept.

Notebook Entry Student Prompt: In your science notebook, write the word diffusion, define it in your own words, and draw **three** examples that illustrate the process of diffusion in everyday situations.

Background
Diffusion is an important concept in all fields of science. The two most common demonstrations in middle school are the diffusion of food coloring in water (liquid in liquid) and diffusion of perfume in air (gas in gas). Other examples of diffusion are solids dissolving in liquid (e.g., making tea or coffee), solids dissolving in air (e.g., smoke or dust particles mixing with air), and gases dissolving into liquid (e.g., oxygen in the air dissolving into lake water).

In biological systems, oxygen diffuses into the mucous lining of our lungs and enters our blood stream and cells. After food has been broken down into molecules by our digestive system, food molecules diffuse into the blood and enter and leave our cells. Likewise, waste molecules move from cells to blood to lungs, kidneys or skin cells by diffusion.

Most people don't know that diffusion can even take place in solids. When metals are heated to high temperatures and formed into alloys (i.e., brass is zinc-copper alloy), solid molecules are able to "vibrate" their way through the other solid molecules.

Lesson Sixteen: Gases and Air Pressure
Student Investigation

Focus Question: How does pressure affect gases?

Materials: Microscale Vacuum Apparatus (Mini Bell Jar and Syringe Set-Up), small plastic vial (10-15 mL), small balloon, thermometer, water, marshmallows and/or aerosol shaving cream

Procedure 1. Bell Jar and Balloon
1. Put together the bell jar apparatus.
2. Remove the bottom of the bell jar.
3. Partially fill the small balloon with air and tie it off.
4. Place the balloon on the bottom of the bell jar and cover it with the bell.
5. Evacuate the air in the bell jar by pumping the syringe. Observe what happens to the balloon.
6. Open the valve and let air flow into the jar. Observe what happens when the air is added.

Observation 1: Record what happened to the balloon when air was evacuated and allowed to flow back into the jar.

Procedure 2. Bell Jar and Beaker with Water
1. Fill a small beaker about ½ full with warm water.
2. Measure the temperature of the water with the thermometer.
3. Place the beaker inside the bell jar and evacuate the air by pumping the syringe. Observe what happens.
4. Open the valve and let air flow into the jar. Observe.
5. Measure the temperature of the water with the thermometer.

Observation 2. Record what happened to the water in the beaker when the air was evacuated. Record the temperature of the water at the end of the procedure.

Procedure 3. Bell Jar and Marshmallow and/or Shaving Cream
Prediction: Before you place a marshmallow inside the jar, predict what you think will happen as you increase and decrease pressure.
1. Place the marshmallow inside the bell jar and evacuate the air by pumping the syringe.
2. Open the valve and let air flow into the jar. Observe what happens when the pressure is increased.

Observation 3: Record what happened to the marshmallow.

Claims and Evidence
1. What is the relationship between gas pressure and volume?
2. What is the relationship between the boiling point of a liquid, air pressure, and temperature?

New Understandings
1. What is a vacuum? Did you create a total or partial vacuum in your bell jar? Explain.
2. How does air pressure change as you go up in altitude? Could this affect breathing?
3. When a weather balloon is released and floats higher into the atmosphere, will the size of the balloon change? Explain.

Teacher Notes: Gases and Air Pressure
Lesson Sixteen: Student Investigation

Rationale: In Lesson Twelve students investigated the effect of temperature on gases at constant pressure. In this lesson, students study the effect of pressure on gases at a constant temperature.

Materials: Microscale Vacuum Apparatus* (Mini-Bell Jar and Syringe Set-Up), small plastic vial (10-15 mL), small balloon, thermometer, water, marshmallows and/or aerosol shaving cream

*Microscale Vacuum Apparatus: This is one of the few investigations that require a specialized item that consists of a small, clear plastic bell jar (8.5 cm or 3.5" high) with base plate, and vacuum pump syringe. It is carried by more than one company and can be ordered online. As of this writing, Educational Innovations (Teacher Source Website) offers the apparatus online for $35.99 + shipping and handling. It comes with items that can be placed in the jar. Depending on your budget, you can either order one set-up per 4-5 students or do the investigation as a demonstration. Of course, a larger, glass bell jar and vacuum pump can also be used, if available.

Teacher's Role: Your role during the investigation is to facilitate lab work by managing the use of supplies, observing student work, and asking probing questions. As you monitor student work, continually assess student understanding as evidenced by their comments, questions, and notebook entries.

Background: At a constant temperature, reducing pressure will increase volume. When the pressure is increased, volume decreases. Pressure is inversely proportional to volume. Boiling water by reducing air pressure is usually a discrepant event for students and difficult for them to understand. For students, the concept of boiling has always been associated with increased temperature rather than reduced pressure. The next reading provides a written explanation; thus, it is important that all students have first-hand observations of the water "boiling" in the bell jar. They should have the opportunity to feel the water and note that it did not change temperature. They should also measure the water's temperature to verify that it stayed the same.

Procedure 1. Bell Jar and Balloon
Observation: As air was evacuated, the partially filled balloon expands. As the air flowed back into the jar, the balloon deflates.

Procedure 2: Bell Jar and Beaker with Water
Observation: The water, after many pumps of the syringe, begins to bubble or boil. The temperature of the water before and after stays the same.

Procedure 3: Bell Jar and Marshmallow and/or shaving cream
Observation: The marshmallow expands as the air is evacuated and then contracts as if it were squished when air flows back into the jar. This is similar for shaving cream.

Claims and Evidence
1. **What is the relationship between gas pressure and volume?** As gas pressure is reduced, volume increases. As gas pressure increases, volume is reduced. As the air was evacuated, air molecules were removed and pressure decreased on the outside of the balloon. The molecules inside the balloon continued to exert pressure and the difference in pressure inside and outside of the balloon caused the balloon to expand.
2. **What is the relationship between the boiling point of a liquid, air pressure and temperature?** As pressure is reduced, liquids boil at lower temperatures. As the pressure was decreased, the liquid began to bubble or boil. Yet, the temperature of the water remained the same. The reduction in air pressure resulted in the phase change, not an increase in temperature.

New Understandings
1. **What is a vacuum?** A vacuum is a space that is without air (gas molecules). It is empty space without matter. **Did you create a total or partial vacuum in your bell jar? Explain.** A partial vacuum. This equipment does not allow the evacuation of all the air.
2. **How does air pressure change as you go up in altitude?** Air pressure decreases as you go up in altitude because the air is thinner or less dense meaning it has fewer air molecules. **Could this affect breathing?** Yes. There are less oxygen molecules to breath.
3. **When a weather balloon is released, will the size of the balloon change? Why or why not?** As a weather balloon increases in altitude, it will expand if it is made of an expandable material. Weather balloons sometimes reach heights of 25 miles above the Earth. They sometimes expand to the point that they disintegrate.

Teaching Tip: Students love this activity. They especially like seeing the marshmallow expand and contract. They will frequently ask to place other objects in the jar. Use your judgment. Aerosol products like shaving cream, packing peanuts, bubble wrap, and other items with air spaces are fun to watch expand and contract.

Reference Notes: The activities in Parts 1 and 2 are modified from two of the activities in the *Microscale Vacuum and Pump Set General Information* packet included in the Microscale Vacuum Apparatus purchased from Flinn Scientific Inc. This packet is eleven pages and was written by James Housley, 1999 – Transparent Devices.

Microscale Apparatus Purchasing Information
Educational Innovations, Inc. 362 Main Avenue, Norwalk, CT 06851 www. teachersource.com

Flinn Scientific Inc., P.O. Box 219, Batavia, Illinois 60510-0219 www.flinnsci.com

Lesson Seventeen: Boiling Hot or Boiling Cold?
Student Reading: Anticipation Guide

I. Read each of the following statements and state whether you (A) agree or (D) disagree by circling the appropriate letter.

A or D   1. A vacuum is empty space with no gas molecules.

A or D   2. A partially filled balloon will contract when air pressure is lowered.

A or D   3. Molecules at the surface of a liquid are constantly entering (condensing) and leaving (evaporating) the liquid creating vapor pressure above the liquid's surface.

A or D   4. It is possible to boil water without raising the temperature of the water.

A or D   5. Water boils at 100°C at sea level.

A or D   6. In the mountains, water will boil at a higher temperature than at sea level.

A or D   7. In the mountains, boiled food takes longer to cook.

A or D   8. Water boils at a lower temperature than cooking oil.

A or D   9. Only pressure affects the behavior of gases.

II. With a partner, discuss your responses. You may change your answers before you do the reading, if you wish.

III. Read the article, "Boiling Hot or Boiling Cold?" As you read, make any corrections to the Agree/Disagree statements. Be prepared to back up your answers with evidence from your reading.

IV. Discuss the statements again with either your partner and/or your teacher.

V. Summarize the major ideas by writing sentences that relate the following terms: vacuum, gas molecules, liquid molecules, air pressure, vapor pressure, sea level, boiling point.

Lesson Seventeen: Boiling Hot or Boiling Cold?
Student Reading

As air is evacuated out of a bell jar, molecules are removed. Because there are fewer molecules, the air pressure inside the jar decreases. If all the air molecules were removed, a perfect **vacuum** would be created. So far, no one has been able to create a perfect vacuum, only a partial vacuum. Outer space most likely comes the closest to a perfect vacuum, but it still has some molecules.

When a partially filled balloon is placed in a bell jar, it increases in volume as the air molecules outside of the balloon are removed. The pressure inside the balloon becomes greater than the pressure outside the balloon and the balloon expands.

The processes of **evaporation** and **condensation** cause the atoms or molecules of the liquid to exert a vapor pressure above the liquid. Vapor pressure will increase when liquids are heated. When the **vapor pressure** equals the atmospheric pressure, liquids boil. At the boiling point, the temperature of the liquid temporarily stops rising and molecules from all parts of the liquid, not just the surface, escape quickly from the liquid into the air.

You can also "boil" a liquid without heating it. All you need to do is reduce the atmospheric vapor pressure to equal the vapor pressure at the surface of the liquid. In the bell jar, the atmospheric pressure was reduced to the point that water boiled at room temperature.

The **boiling point** of a liquid is a **characteristic property**. Because the boiling point of a liquid depends on atmospheric pressure, boiling points are standardized at **sea level pressure and at 25°C**. The boiling point of water is 100°C (212°F) at sea level. Vinegar boils at a higher temperature, approximately 118°C (244°F) at sea level. Alcohol boils at a lower temperature, 82.4°C (180°F). Mineral oil molecules require even more kinetic energy than water, vinegar, or alcohol to escape from the liquid to the vapor phase. Because mineral oil is heavy and viscous, it must reach at least 260°C (500°F) before boiling.

Teacher Notes: Boiling Hot or Boiling Cold?
Lesson Seventeen: Student Reading

Rationale: The reading builds on the student's previous observations and provides a model of a written explanation for the relationship between vapor pressure and boiling point.

Strategy: Anticipation Guide. See Literacy Resource Guide.
    Before Reading: Lesson Sixteen: Gases and Air Pressure, Anticipation Guide I & II.
    During Reading: Anticipation Guide Part III.
    After Reading: Anticipation Guide Parts IV & V.

Materials: Student copies of Anticipation Guide and Reading: Boiling Hot or Boiling Cold?

Teacher's Key: Anticipation Guide-Boiling Hot or Boiling Cold?
I. Read each of the following statements and determines if you (A) agree or (D) disagree.
1. Agree. **A vacuum is empty space without gas molecules.** Agree, if referring to a perfect vacuum.
2. Disagree. **A partially filled balloon will contract when air pressure is lowered.** The balloon will expand.
3. Agree. **Molecules at the surface of a liquid are constantly entering (condensing) and leaving (evaporating) the liquid creating vapor pressure above the liquid's surface.**
4. Agree. **It is possible to boil water without raising the temperature of the water.** Just lower the vapor pressure.
5. Agree/Disagree. **Water boils at 100°C.** Agree, if students qualify statement to indicate that water must be at sea level. Disagree, if students refer to boiling water at altitude.
6. Disagree. **In the mountains, water will boil at a higher temperature than at sea level.** When the air pressure is lower, the vapor pressure and air pressure equalize at a lower temperature. Thus, boiling happens at a lower temperature. Note: At very high elevations, water may not boil hot enough to purify water!
7. Agree. **In the mountains, boiled food takes longer to cook.** Since water boils at a lower temperature, food takes longer to cook.
8. Agree. **Water boils at a lower temperature than cooking oil.** Cooking oil molecules, like mineral oil, require more kinetic energy due to the stronger attractive forces between their molecules.
9. Disagree. **Only pressure affects the behavior of gases.** Both temperature and pressure affect the behavior of gases.

V. Summarize the major ideas by relating the following terms: vacuum, gas, liquid, molecules, air pressure, vapor pressure, sea level, boiling point.
Sample Response: A **vacuum** is a space without **molecules**. Moving **gas molecules** cause **air pressure**. **Boiling point** depends on **air pressure and vapor pressure (which is dependent on temperature)**. **Boiling points** are reported at standard pressure at **sea level** and at 25°C.

Lesson Eighteen: Can Crush
Part 1. Teacher Demonstration

How to Perform the Can Crush Demonstration

Before handing out the student page, perform the demonstration described below. Make sure everyone can see what you are doing. Ask students to watch closely. Don't tell them that you are crushing a can. You will repeat this demonstration after a class discussion.

Materials: Safety goggles for you and your students, hot plate, empty aluminum soda cans*, water and measuring cup for water, small plastic tub with ice water, tongs, pot holder, thermometer. *Since you will repeat this demonstration a number of times, have at least 5 cans ready for crushing per class.

To Do

1. Plug in the hot plate.
2. Add water to the tub and add ice.
3. Place the thermometer into the tub and record the temperature.
4. Add about 50 mL of water to the pop can.
5. Take the temperature of the water in the pop can and record.
6. Place the pop can on the hot plate.
7. Heat the can until you see steam coming from the top of the can. **Note: Steam is not water vapor. Steam is tiny liquid water droplets. You cannot see water vapor.** However, when the steam appears you can be sure that enough water vapor is present to move to the next step.
8. With the tongs, grab the can, turn it **upside down, and submerse it** into the cold water so that the can opening goes into the water first.
9. The can will crush and make a great sound!
10. Carefully, remove the can and lift it high above the tub so that the students can see water running out of the can. If you do this just right, it should be obvious that much more water is coming out of the can than you put in.
11. Without further discussion, hand out the student page.

Lesson Eighteen: Can Crush
Part 2. Student Investigation

Introduction: Sometimes science seems like magic. You have just watched your teacher crush a pop can with very little effort. Let's take another look, and try to explain why in scientific terms.

Focus Question: What caused the can to crush?

Initial Observations/Ideas: Before your teacher crushes another can, write down what you observed and any other thoughts/ideas about the can crush.

More Observations: Your teacher will crush another can or two. This time your teacher will describe what's being done and provide more observations/data for you to record in your science notebook.

Claims and Evidence
1. Based on your additional observations, what can you claim caused the crushing of the can?
2. Did your additional observations change your earlier explanation?

New Understandings
1. What happened to the pressure inside the can when the water vapor condensed? Explain.
2. Give an example of an inference you made based on an observation. Tell the observation and the inference.
3. Would the can crush if it had not been turned upside down when it entered the water? Explain.

Reflections: Did the demonstration help you understand why the can crushed?

Teacher Notes: Can Crush
Lesson Eighteen: Part 2. Student Investigation

Rationale: This high interest demonstration with student participation integrates and applies a number of basic concepts from this lesson set. It increases student motivation, provides an ongoing assessment opportunity, applies literacy skills (listening, speaking, writing), and teaches complex concepts. Students frequently watch this classic demonstration and come away without the opportunity to learn the science behind the crush. By placing this extended demonstration at this point in the lesson set, students will have a better chance of understanding the complexities of the concepts involved.

Materials: Student copies of Student Investigation: Can Crush – Part 2 and can crush demonstration materials

Teacher's Role: Your role is to facilitate class discussion by clarifying and listing student observations, asking probing questions, generating student thinking, and summarizing key concepts and ideas.

Before you repeat the first demonstration, make sure students write down their initial observations and ideas. As you repeat the demonstration, use your questioning skills to generate student responses. Draw their attention to key observations by asking questions like the ones below.
- Does it matter how much water I put in the can?
- What happens to the water in the can on the hot plate?
- What phase change is taking place as I heat the water on the hot plate?
- Will you get the same results if the tub water is warm? Why or why not?
- Did I put the can in the tub right side up or upside down? Does it make a difference?
- What happens to the water vapor when the can is submerged in water?
- What happens to the volume of a gas when the gas condenses?
- Is air pressure the same or different inside and outside the can when the can is on the hot plate?
- What is a closed system? What is an open system?
- What happens to the air pressure inside the can when it is submerged in water?
- How is what happened to the marshmallow in the bell jar similar to what's happening here?

Have students write down more observations as you repeat the demonstration a few more times. During one repeat, catch the water coming out of the can, measure it, and compare it to the amount of water you put in.

Background: The Chart below summarizes key observations and data.

Can Crush Observations and Inferences	
Before Heating	**Observations/Inferences**
Amount of water in can	50 mL
Temperature of tub water with ice	10°C (approximate)
Temperature can water	27°C (approximate)
After Heating	
Temperature can water	100°C
Phase Change of can water on hot plate	Boiling-Liquid to Vapor
Pressure on outside of can on hot plate	Atmospheric
Pressure on inside of can on hot plate	Atmospheric-Open System
Phase Change of can water as it enters the Tub	Condensation-Vapor to Liquid
Pressure on inside of can as it enters the Tub	Less than atmospheric-Closed System
Pressure on outside of can as it enters the Tub	Atmospheric
Amount of water in can after crushed	100 mL (approximate – will be more than amount added)

Claims and Evidence

1. **Based on your additional observations, what can you claim caused the crushing of the can?** The **difference in air pressure** inside the can and outside the can after it entered the tub caused the can to crush. **Evidence & Explanation:** Before the can of hot water entered the tub, the pressure on the inside and outside of the can was about the same. As the can entered the ice water upside down, the water vapor inside the can condensed reducing the pressure inside. As water vapor phased into a liquid, the volume of the water vapor was reduced and water rushed into the can. We know this because when we measured the water inside the crushed can, it was greater than the water we added to the uncrushed can. But, the water could not enter fast enough to equalize the pressure between the inside and outside of can. The outside atmospheric pressure crushed the can.

New Understandings

1. **What happened to the pressure inside the can when the water vapor condensed? Explain.** The pressure inside the can decreased. When water vapor condenses or cools, the amount of water vapor decreases. With fewer molecules in the vapor phase, the pressure inside the can decreases. Note: Because the can entered the water upside down, the water outside the can could not rush in fast enough to equalize the pressure difference as the water condensed. If the can enters the water right side up, the air can enter fast enough and the can won't crush.

2. **Give an example of an inference you made based on an observation. Tell the observation and the inference**. Inferences are statements that are not supported by direct observation. In this investigation, we are inferring that the can is surrounded by atmospheric pressure. We did not measure the pressure. We do not have to infer the amount of water we put in the can, the temperature of the water, or the amount of water that came out of the can. These we were able to measure. We also had to infer that the condensing water vapor created a region of lower pressure inside the can since we did not measure the pressure inside the can.

3. **Would the can crush if it had not been turned upside down when it entered the water? Explain.** No. If the opening of the can was not submerged, air could have rushed into the can to equalize the pressure. When the can is turned over, water can't rush in fast enough to equalize the pressure, and the can collapses.

Teaching Tips

1. Students love to take part in this demonstration. Be prepared to have as many students crush a can as possible – the more, the better!

2. One teacher used this demonstration as a reward for good behavior and completing class assignments, like a reading. When students finished early, they were allowed to crush more cans. It worked great.

3. A few days before the demo, have students collect cans and store them in your room.

4. Remember to recycle all the crushed cans.

Lesson Nineteen: Graphing Phase Changes
Student Chart Reading, Graphing, and Data Interpretation

Focus Question: What does analyzing data reveal about phase changes?

Boiling point is a characteristic property of matter. To find the boiling point, a chemist heated each of the liquids below and recorded the liquid's temperature every minute.

Temperature of Liquids Over Twelve Minutes of Heating (sea level)			
Time In Minutes	Water Temperature (°C)	Isopropyl Alcohol Temperature (°C)	Oxetane Temperature (°C)
0	25	25	25
1	28	30	31
2	35	42	40
3	42	51	50
4	51	60	50
5	69	77	50
6	78	83	50
7	85	88	50
8	95	88	51
9	99	88	52
10	100	88	53
11	100	88	56
12	100	90	58

Graphing the Data: In your science notebook, make a graph of the data.

Data Interpretation
1. Describe the shape of the graph at the boiling point of water. Hint: What have you learned about the boiling point of water in previous readings?
2. Based on the shape of the boiling point graph for alcohol, what might be its boiling point?
3. What is the boiling point of oxetane? State your evidence.
4. Which substance has the highest boiling point?
5. Which substance has the lowest boiling point?

New Understandings
1. The temperature of water reached a "plateau" at 100°C. Explain.
2. Would 10 mL of water boil at the same temperature as 1000 mL of water? Explain.
3. If you were camping in the mountains at an altitude of 5000 feet above sea level, will water boil at the same temperature as sea level? Explain your thinking.
4. Oxygen is a gas at room temperature and sea level pressure. Does it have a boiling point?
5. A mixture of water, alcohol, and oxetane is prepared at room temperature (25°C). If the mixture is heated in a laboratory to 65°C, will the composition of the mixture change?

Synergized Middle School Chemistry

Going Further: Freezing point is also a characteristic property of matter. To find the freezing point, a chemist cooled each of the liquids and recorded the temperature every minute.

Temperature of Liquids Over Twelve Minutes of Cooling (Sea Level)			
Time In Minutes	Water Temperature (°C)	Isopropyl Alcohol Temperature (°C)	Oxetane Temperature (°C)
0	25	25	25
1	22	15	12
2	18	5	2
3	12	0	-10
4	9	-16	-25
5	5	-20	-40
6	3	-30	-50
7	1	-50	-54
8	0	-65	-61
9	0	-75	-66
10	0	-88	-70
11	0	-88	-76
12	-3	-88	-80

Graphing the Data: In your science notebook, make a graph of the data.

Interpreting Data
1. Describe the shape of the graph at the freezing point of water. Hint: What have you learned about the freezing point of water in previous readings?
2. Based on the shape of the freezing point graph for alcohol, what might be its freezing point?
3. What is the freezing point of oxetane? State your evidence.
4. At what temperatures is water a liquid?
5. At what temperatures is alcohol a liquid?
6. At what temperatures is oxetane a liquid?

New Understandings
1. As a sample of water is cooled, the temperature of water reached 0°C and then went lower. Explain.
2. Would a drop of water freeze at the same temperature as a cup of water?
3. Oxygen is a gas at room temperature and sea level pressure. Does it have a freezing point?
4. What new questions do you have about phase changes?
5. Did graphing the data help you interpret the data?

Reflections: Other questions or comments?

78

Teacher's Notes: Graphing Phase Changes
Lesson Eighteen: Student Investigation

Rationale: The phases of matter are solid, liquid, and gas. Phase changes result from changes in temperature. This lesson integrates previous concepts with graphing and data interpretation skills. Reading and interpreting charts and graphs is a fundamental mathematics and science skill that requires repeated learning and practice opportunities.

Materials: Student pages, graph paper, calculator optional, computer optional.

Teacher's Role
Your role is to teach students how to construct a graph of data. Student graphs should include: title, axes labels with measurement, appropriate scale for range of data, and line graph connecting accurate data points. Some students may need more graphing instructions than others. Even though some students may be able to determine the boiling points from the data alone, knowing what a boiling point graph looks like is critical to future understanding of phase changes.

Background
Isopropyl alcohol is the main ingredient in rubbing alcohol. Oxetane is a component in some cancer drugs. CAUTION: Isopropyl alcohol and oxetane should never be heated in the classroom. Students need to know that experimenting with chemicals is dangerous, unless under the careful supervision of a teacher or chemist.

When liquids reach their boiling point, the temperature remains constant because the average kinetic energy of the molecules remains the same. However, the molecules continue to absorb heat energy. The point where temperature remains constant is the boiling point. It is also the point where the vapor pressure equals the atmospheric pressure.

When liquids freeze, heat energy is absorbed by the surroundings. The temperature begins to fall as the kinetic energy of the molecules decreases. At the freezing point, heat is released at the same rate that it is being absorbed from the surroundings. Thus, the temperature plateaus until heat is released faster than it can be absorbed allowing the solid to cool below the freezing point.

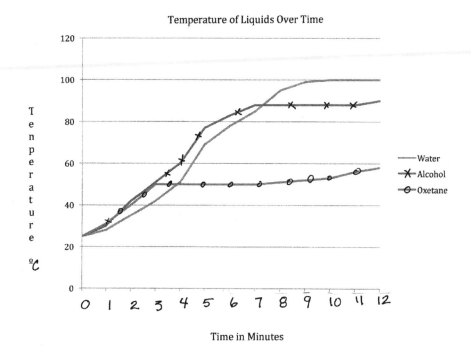

Temperature of Liquids Over Time

Interpreting Data
1. **Describe the shape of the graph at the boiling point of water. Hint: What have you learned about the boiling point of water in previous readings?** The line on the graph levels out or plateaus at 100°C. This is the boiling point of water at sea level.
2. **Based on the shape of the boiling point graph for alcohol, what might be its boiling point?** The rubbing alcohol graph has a plateau at 88°C. This is the boiling point of rubbing alcohol. It is the point the temperature remained constant.
3. **What is the boiling point of oxetane? State your evidence.** 50°C. It is the point the temperature remained constant.
4. **Which substance has the highest boiling point?** Water.
5. **Which substance has the lowest boiling point?** Oxetane.

New Understandings
1. **The temperature of water reached a "plateau" at 100°C. Explain.** When the vapor pressure equals the atmospheric pressure, the liquid boils. When a phase change happens, heat energy is either absorbed or emitted without changing temperature. In this case, heat is absorbed until all the molecules have gained enough heat energy to escape into the gas phase. Then, the temperature will rise.
2. **Would 10 mL of water boil at the same temperature as 1000 mL of water? Explain your thinking.** Yes. It will just take more time and heat energy. Boiling point is not dependent on volume or the amount of substance.

3. **If you were camping in the mountains at an altitude of 5000 feet above sea level, will water boil at the same temperature as sea level? Explain your thinking.** No. Water will boil at a lower temperature because the atmospheric pressure is lower at altitude.

4. **Oxygen is a gas at room temperature and sea level pressure. Does it have a boiling point?** Yes. It would be lower than room temperature. It boils at -182.95°C (-297.31°F).

5. **A mixture of water, alcohol, and oxetane is prepared at room temperature (25°C). If the mixture is heated in a laboratory to 65°C, will the composition of the mixture change?** This is an example of distillation of liquids. The oxetane will boil off at 50°C leaving the water and alcohol behind.

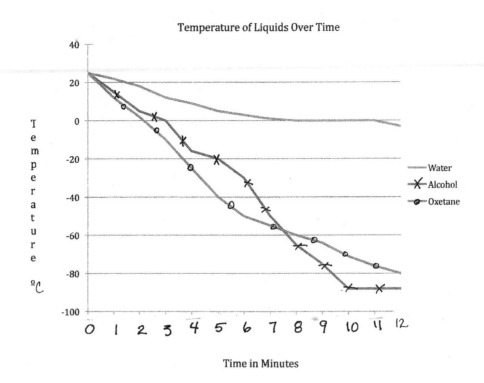

Temperature of Liquids Over Time

Time in Minutes

Interpreting Data

1. **Describe the shape of the graph at the freezing point of water. Hint: What have you learned about the freezing point of water in previous readings?** The line on the graph levels out or plateaus at 0°C. This is the freezing point of water.

2. **Based on the shape of the freezing point graph for alcohol, what might be its freezing point?** The line on the graph plateaus at -88°C. This is the freezing point of alcohol.

3. **What is the freezing point of oxetane? State your evidence.** Unknown. The freezing point was not reached because the temperature continues to fall.

4. **At what temperatures is water a liquid?** Between 0°C and 100°C at sea level.

5. **At what temperatures is alcohol a liquid?** Between -88°C and 88°C at sea level

6. **At what temperatures is oxetane a liquid?** -80°C or lower to 50 °C at sea level.

New Understandings

1. **As a sample of water is cooled, the temperature of water reached 0°C and then went lower. Explain.** When molecules lose kinetic energy, they slow down and the liquid phases into a solid. When a phase change happens, heat energy is either absorbed or emitted without changing the temperature. When a solid cools further, heat energy is emitted at a greater rate than the heat energy is being absorbed from the surroundings.

2. **Would a drop of water freeze at the same temperature as a cup of water? Yes.** Freezing point is not dependent on the amount or volume of the water. However, it will take longer to freeze a cup of water because more heat must be emitted.

3. **Oxygen is a gas at room temperature and sea level pressure. Does it have a freezing point?**
    Yes. All substances go through phase changes. Oxygen freezes at -218.79°C (-361.82°F). It boils at -182.95°C (-297.31°F). It is a liquid between -218.79°C and -182.95°C.

Teaching Tips
1. This lesson can be adapted to the graphing skill level of your class. For younger learners, you can have them graph just the boiling point of water. Students are intrigued by the boiling point plateau. Some students may need to make three separate graphs rather than put three graphs on the same grid. Three lines on one graph can be confusing to many students.

2. The freezing point graph can be given as a homework assignment or optional assignment.

Lesson Twenty: Reviewing Matter's Phases
Student Review

Part 1. Word Sort

Understanding the phases of matter and phase changes are basic to understanding chemistry. Word sorts help make meaning out of the many new words and ideas. But, before beginning the word sort, read the paragraphs below to learn about sublimation and deposition.

Sublimation and Deposition

Melting, freezing, evaporating, and condensing are the most commonly known phase changes. Two additional phase changes are **sublimation** and **deposition**.

Sublimation is the process of going directly from a solid to a gas. An example is a deodorant stick that changes directly from solid to gas. You may also know about dry ice. Dry ice is solid carbon dioxide. It is used to keep food cold or to create a spooky effect on Halloween. Dry ice sublimes and changes from a solid to a gas at normal room temperature without going through a liquid phase.

Deposition is the process of going from a gas to a solid. The most familiar example is when frost appears on a window. Water vapor can phase directly to ice and create frost when the temperature and air pressure are just right.

Materials: Word Sort Envelope

To Do

1. Working with a partner, place the word cards on the table.
2. Organize the cards in a way that shows the relationships between the words.
3. After sharing your sort with the class, draw a word sort concept map in your notebook.

Student Word Sort Cards: Master Sheet – Copy, Cut Out, and Place in Envelope

Crystal Lattice	Heating to Melting	Fixed Shape and Volume	Viscous
Pourable	Heating to Vaporization	Fixed Volume but Takes the Shape of the Container	No Fixed Shape or Volume
Molecules vibrate	Attractive Forces Hold molecules Together	Molecules Vibrate, Rotate, spin, and move around each other	Attractive Forces Hold Molecules Together
Molecules Vibrate, Rotate, spin, and move around each other	Attractive forces do not hold molecules together	Collisions give rise to air pressure	Melts
Vapor/Gas	Sublimes- Changes from Solid to Gas	Deposits- Changes from Gas to Solid	Condenses
Freezes	Evaporates (Vaporization)	Liquid	Cooling to Condensation
Solid	Water	Ice	Water vapor

Part 2. Student Vocabulary Review: First Word - Last Word

Directions: Select one of the words listed at the bottom and write the word vertically in your notebook, as shown below for the word "matter".

M
A
T
T
E
R

Next, write one word or short phrase related to our study of phase changes that starts with each letter horizontally, as shown in the example below.

Mass
Atoms
Temperature
Time
Energy
Reduce

Finally, write one or more sentences that use each of the words. Underline the words. An example is provided below.

> Matter takes up space and has mass. Matter is made of atoms or molecules. When the temperature of matter increases over time, the atoms or molecules gain energy. If you reduce the temperature, the atoms or molecules will lose energy.

Words to Select: SOLIDS, LIQUIDS, VAPORS, PHASES, EVAPORATION, CONDENSATION, BOILING, PRESSURE, TEMPERATURE, CRYSTAL

Part 3. Revisiting Everyday Chemicals

On Your Own (10 min)
1. In your notebook, make three columns entitled solids, liquids, and gases. Write as many substances as you can in the time provided under each column. Make sure you have at least five solids, five liquids, and five gases.
2. Pick one solid, one liquid, and one gas and describe it in as many ways as you can using single words or short phrases.

Pair-Share (10 min)
1. Without telling your partner the name of your everyday chemicals, read just your descriptive words or phrases and see if your partner can guess your everyday chemicals.
2. After you have both described your everyday chemical, compare your chemical lists and see how many chemicals you named in common.

Team Work (20 min) Revisiting Lesson One
1. Review the Everyday Chemical list made in Lesson One of this unit. What chemicals would you add or subtract from the list and why?
2. You classified chemicals as solids, liquids, and gases. What do you know now that might change your classification?
3. Review your Lesson One notebook entry. Find the questions that you asked about classifying the chemicals. What questions can you answer now and which questions still need answers?

New Understandings
1. What have you learned about everyday chemicals that you didn't know before?
2. What have you learned about chemical safety?

Teacher Notes: Reviewing Matter's Phases
Lesson Twenty: Student Review

Part 1. Word Sort and Concept Map

Rationale: Before taking the post-assessment students have the opportunity to revisit concepts and apply their learning. Word sorts are a kinesthetic way to engage students. Sorting words and then writing them motivates reluctant learners. Word sorts are best done after students have used the words in the context of an investigation and in a reading.

Materials & Preparation: Copy, cut and place the words on the Master Sheet in an envelope or small baggie. **The Master Sheet is found after the student directions.** Make enough word sets for every group in the class. To prevent words from being lost and to facilitate clean up, number each of the word sets and their envelope. For example, place a "1" on each word and a "1" on envelope one. Thus, all the "1" words go in the "1" envelope and "2" words in the "2" envelope, etc. Make extra sets, especially if multiple classes will use the words.

Teacher's Role: Provide each group with a set of words and allow them time to sort the words in different ways. As you monitor the activity, ask students to explain their word arrangements. Allow them to use their science notebook and other resources during the sorting. When the groups are finished, have groups walk around the room to see how other groups sorted the words. As long as the words are related correctly, any arrangement is acceptable. When students are finished discussing their sorts, have each student draw their word sort concept map in their science notebook.

Background: Students should display the words to show the following relationships.
Solid: crystal lattice, heating to melting, fixed shape and volume, molecules vibrate, attractive forces hold molecules together, melts, sublimes, ice
Liquid: viscous, pourable, heating to vaporization, fixed volume but takes the shape of the container, molecules rotate, spin and move around each other, attractive forces hold molecules together, freezes, evaporates, water
Gas/vapor: no fixed shape or volume, attractive forces do not hold molecules together, collisions give rise to air pressure, molecules rotate, spin and move around each other, deposits, condenses, vapor, cooling to condensation, water vapor

Part 2. First Word-Last Word

Materials: Copies of student page, science notebook, and other appropriate resources

Teacher's Role: If the students have not used this strategy before, model the process explained on the sheet with the word MATTER. Have students select the words as you model.

Teaching Tip: First Word-Last Word can be used any time during a lesson set to reinforce vocabulary terms and does not have to wait for a review lesson.

Part 3. Revisiting Everyday Chemicals (See Lesson One – Pre-assessment)

Rationale: Revisiting a previous lesson allows students an opportunity to recognize what they have learned, correct misconceptions, and ask new questions. Revisiting Lesson One: Everyday Chemicals makes an excellent focus for a final discussion before students take the post-assessment.

Materials: Student science notebooks, team chemical lists from lesson one

Teacher's Role: Your role is to facilitate a discussion of everyday chemicals in order to review and reinforce key concepts. Use the students' work and their questions to begin the discussion.

Lesson Twenty-One: Matter's Phases
Student Post-Assessment

Part 1. Multiple Choice

Directions: Write the letter of the "best" answer in the space provided. If you wish to explain why you chose that answer, write a response by the question.

___1. When sugar is added to water, it
    a. melts into the liquid.
    b. forms a gas.
    c. changes to a solid.
    d. dissolves into the liquid.

___2. The kinetic energy of atoms or molecules is greatest in
    a. solids.
    b. liquids.
    c. gases.
    d. crystals.

___3. When matter changes phase from solid to liquid, which of the following occur?
    a. Molecules gain energy and all attractive forces are overcome.
    b. Molecules gain energy but attractive forces still hold them together.
    c. Molecules lose energy and all attractive forces are overcome.
    d. Molecules lose energy but attractive forces still hold them together.

___4. Which best describes the movement of molecules in a liquid?
    a. Molecules vibrate and hold their position.
    b. Molecules spin and hold their position.
    c. Molecules move around with little or no attraction to other particles.
    d. Molecules vibrate, spin, and move around but remain attracted to other molecules.

_____5. Which substance listed in Table 1 is a liquid a 27°C?

Melting and Boiling Points of Four Substances		
Substance	Melting Point	Boiling Point
A	28°C	140°C
B	-10°C	25°C
C	20°C	140°C
D	-90°C	14°C

    a. Substance A
    b. Substance B
    c. Substance C
    d. Substance D

_____6. When a bottle of perfume is opened in the corner of a room, the perfume molecules will
    a. stay in the bottle.
    b. diffuse throughout the room.
    c. evaporate from the bottle but stay within a few feet of the bottle.
    d. float to the ceiling.

_____7. Water boils at 100°C at sea level. At higher altitudes, at what temperature will water boil?
    a. less than 100°C
    b. the same, 100°C
    c. greater than 100°C
    d. stay at 100°C then get cooler.

_____8. If all the molecules were removed from a piece of paper, what would be left?
    a. The paper would still be there but it would have less mass.
    b. The paper would be exactly the same.
    c. There would be no paper left.
    d. There would be ashes left.

_____9. A student finds an unlabeled white solid. The student thinks the solid is table salt. Which of the following tests would provide the *best evidence* to support the student's claim?
    a. Dissolve the substance in water.
    b. Look at the size of the particles.
    c. Grind the solid into a powder.
    d. Look at the shape of the crystals.

_____10. A balloon is filled with air. It is placed in a freezer. The balloon will
   a. decrease in size.
   b. stay the same size.
   c. increase in size.
   d. decrease then increase in size.

_____11. When ice absorbs heat, it will
   a. Gain kinetic energy and solidify.
   b. Gain kinetic energy and liquefy.
   c. Lose kinetic energy and vaporize.
   d. Lose kinetic energy and sublime.

_____12. When water vapor in a balloon condenses, the balloon will
   a. increase in size.
   b. stay the same size.
   c. decrease in size.
   d. increase then decrease in size.

_____13. Which of the following is not a characteristic property of matter?
   a. Diffusion
   b. Boiling point
   c. Solubility
   d. Crystal shape

_____14. When astronauts travel to the Moon, what happens to their mass?
   a. Mass increases.
   b. Mass stays the same.
   c. Mass decreases.
   d. Mass cannot be measured on the Moon.

_____15. If you could see a single molecule of water, it would look
   a. Different than a molecule of water vapor.
   b. The same as a molecule of ice.
   c. The same as a molecule of mineral oil.
   d. Different from other molecules of water.

Part 2. Match and Write
Directions: Below are ten numbered words used during our study of matter's phases.
1. Match each numbered word with the best matching lettered phrase at the bottom of the page. You can use each letter only once.
2. Place the letter in front of the word's number.
3. Write a sentence using both the numbered word and lettered phrase in the space provided by the word. Underline the numbered word and the lettered phrase in your sentence. Note: you may change word form. For example, "evaporate" can be changed to evaporation, evaporating, or evaporates, if needed.

___1. Solid

___2. Liquid

___3. Gas

___4. Kinetic

___5. Evaporate

___6. Matter

___7. Diffusion

___8. Freeze

___9. Crystallization

___10. Temperature

Words & Phrases to Match to Numbered Words Above

A. Phase change from liquid to solid	F. Energy of motion
B. Phase change at liquid's surface	G. Characteristic property of solids
C. Definite volume but not shape	H. Affects phase changes
D. Takes up space and has mass	I. Random mixing of molecules
E. No definite shape or volume	J. Definite shape and volume

Teacher Notes: Matter's Phases
Lesson Twenty-One: Student Post-Assessment

Part 1. Multiple Choice (15 points)
Best Responses:
1 = D, 2 = C, 3 = B, 4 = D, 5 = C, 6 = B, 7 = A, 8 = C, 9 = D, 10 = A
11 = B, 12 = C, 13 = A, 14 = B, 15 = B

Part 2. Match and Write (20 points)
Scoring: 2 points for each question— 1 point for matching a correct phrase and 1 point for making a valid sentence connection between the word and phrase. Suggested responses.

(J) 1. Solid – A **solid** has a **definite shape and volume.**

(C) 2. Liquid – A **liquid** has a **definite volume but not shape.**

(E) 3. Gas – A **gas** does not have a **definite shape or volume.**

(F) 4. Kinetic – **Kinetic** energy is **energy of motion.**

(B) 5. Evaporate – **Evaporation** is a **phase change at a liquid's surface.**

(D) 6. Matter – **Matter** is **anything that takes up space and has mass.**

(I) 7. Diffusion – **Diffusion** is the **random mixing of molecules of liquids and/or gases.**

(A) 8. Freeze – **Freezing** is a **phase change from a liquid to a solid.**

(G) 9. Crystallization – **Crystallization** is a **characteristic property of solids.**

(H) 10. Temperature – **Temperature** affects phase changes.**

Evaluation Suggestion
    Advanced: 30-35
    Proficient: 20-29
    Partially Proficient: 15-19
    Unsatisfactory: 0-14

Lesson Set Two:
Density

OVERVIEW

<u>Lesson Set Concepts</u>

2.1 **Density** is a characteristic property of solids, liquids, and gases.

2.2 **Density** is the **mass** per unit **volume** of a substance and is expressed in g/cm^3.

2.3 **Density** can be calculated for solids, liquids, and gases.

2.4 **Density** of irregular solids can be determined by water **displacement.**

2.5 The **density** value of water changes with temperature.

2.6 Ice floats on water because its **density** is less than the **density** of water.

2.7 The **density** of a liquid will determine its position in a column of liquids with different densities.

Key Vocabulary: density, mass, volume, displacement

Lab Equipment and Supplies: Density		
General Science Equipment & Supplies Per Class: 2-3 Electronic Balances Beakers, 250 mL Graduated cylinders: 25, 50 and 100 mL Pipettes (Beral-type), or eyedroppers Per Student: Science Notebooks Safety Goggles Calculators Metric rulers Graph Paper Permanent marking pens	**Grocery Store: Paper, Plastics, Specialty** Trays for group materials Modeling Clay (different colors) Ice cubes made from water with blue food coloring Paper Towels Spoons, plastic Cups, clear plastic, 5-8 oz.	**Grocery/Box Stores/Pharmacy: Chemical Supplies** Table salt Vegetable oil Salt water solutions - Prepared as directed Grapes Food Coloring Rubbing Alcohol Corn syrup (Karo)

Lesson Set Guide: Density		
Lesson	**Title**	**What Students Do**
1 Intro Assess	Can You Explain Why? Pre-assessment	Students brainstorm reasons to explain why some objects float while other objects sink in water.
2 Invest	The Floating Grape	Students determine the conditions for a grape to float, sink and "hang" in salt and fresh water.
3 Read	Defining Density	Students read about density and begin constructing a semantic map with information from the reading.
4 Invest	Calculating Density	Students compute the density of tap water, stock salt water, team prepared salt water, and grape. Reading provides background for students to determine the density of irregular shaped solids.
5 Invest	Does Density Change?	Students investigate how the amount and shape of a substance affect its density. They determine the density of water by comparing the density of 4 different volumes. They determine the density of different shapes and sizes of modeling clay.
6 Graph	Making Sense of Density Data	Students compare and contrast the densities of metals. Students graph water density values at different temperatures.
7 Invest	Issues With Ice	Students observe and compare the density of ice, water, and oil.
8 Invest	Density to the Rescue	Students apply their knowledge of density by calculating the density of 4 liquids (3 of which are potential water pollutants). Students predict how the liquids will layer in a column based on density and test their predictions.
9 Review	Review	Students complete their semantic map (started in lesson three) as a review of the lesson set.
10 Assess	Density Post Test	Students demonstrate their understanding of density through formal assessment and evaluation.

		Before and During the Lesson		End of Lesson
Assessment Guide - Density				
Lesson	Title	Uncovers Student Ideas	Checks for new understandings	Evaluates learning
1 Assess	Can You Explain Why? Pre-assessment	Can You Explain Why scenarios	Can You Explain Why scenarios	Not scored at this time for accuracy
2 Invest	The Floating Grape	Focus Questions	Claims and Evidence (C&E) 1	New Understandings (NU) 1,2
3 Read	Defining Density	Class Discussion	QAR 1-4	Semantic Map (in progress)
4 Invest	Calculating Density	Focus Questions Predictions	C&E 1, 2	NU 1,4
5 Invest	Does Density Change?	Focus Questions Predictions	C&E 1,2,3 NU 5,6	NU: 1,2,5
6 Graph	Making Sense of Density Data	Class Discussion	Chart reading: 4,6 Graphing : 1,3,4,8	Complete and accurate graph Correct answers to all questions
7 Invest.	Issues With Ice	Focus Question	Observations C&E 1,2,3	NU 1,2
8 Invest.	Density to the Rescue	Focus Questions Predictions	Investigation Plan	Accurate density calculations Successful density column Clean up strategies NU 1-4
9 Review	Review	Semantic Map	Semantic Map	Completed Semantic Map
10 Assess	Density			Part 1: Multiple Choice Part 2: Agree-disagree rewrite

Lesson One: Can You Explain Why?
Student Introduction and Pre-Assessment

The world's largest supertanker, the *Seawise Giant,* was built in 1979. It was built to safely carry a payload measuring 564,763 deadweight tons (the sum of the weights of cargo, fuel, fresh water, ballast water, provisions, passengers, and crew). Weight is generally measured in metric tons (1000 kilograms or 2200 pounds). The overall length of the *Seawise Giant* was 458.45 meters (1,504.1 ft), longer than the height of many of the world's tallest buildings. The boat had 46 tanks, 31,541 square meters of deck, and was too large to pass through the English Channel. The rudder weighed 230 tons and the propeller weighed 50 tons. The *Seawise Giant's* history includes being bombed by Iraqi jets, erupting into flames, sinking in shallow waters off the coast of Iran, and finally being rescued for salvage.

How was it possible that a ship with this much mass was able to float in water?

As you begin this lesson set, think about what you already know about sinking and floating. **On Your Own**: Write and/or draw your responses to the following questions in your science notebook.

Situation 1: Why do icebergs float on top of ocean water?

Situation 2: Salad dressings made with vinegar and oil should be shaken before using. Describe the dressing before you shake it. Why does it look this way?

Situation 3: An iron nail sinks to the bottom of a jar of water, but a boat made of iron can float on top of water. How is this possible?

Pair-Share: Share your ideas and drawings with a partner. Add any additional comments.

Teacher Notes: Can You Explain Why?
Lesson One: Student Introduction and Pre-Assessment

Rationale: This activity identifies student preconceptions about density.

Materials: Classroom Set of Pre-Assessment Student Page. Note: This page is a facilitation guide for student work. Students may not need their own copy.

Teacher's Role: Your role is to facilitate a collaborative learning environment that encourages rich student-to-student discussion about sinking and floating objects. The term "density" is purposely not defined so that you can uncover student ideas as they form their explanations.

Background Information: The activity assesses the students' current understanding of density. Even though students may have experienced elementary science lessons on sinking and floating, middle school students may not have been introduced to the concept of density. The *Seawise Giant* scenario sets the stage for students to begin thinking about what makes matter sink or float. This activity activates their prior knowledge and generates new questions.

Can You Explain Why? See Literacy Resource Section.

Given below are appropriate responses to each situation. These are for teacher background only. The responses should not be shared with students at this time or scored for accuracy. They are used to determine current understanding. Later in the lessons, each situation is revisited.

Situation 1: Why do icebergs float on top of ocean water? Icebergs are made of fresh water. The density of an iceberg is .92 g/cm^3. The ocean (salt water) has an approximate density of 1.02 g/cm^3. The iceberg is less dense than the salt water and therefore floats. Students may note that ice floats on fresh water too. Fresh water's density is 1.0 g/cm^3.

Situation 2: Salad dressings made with vinegar and oil should be shaken before using. Describe the pre-shaken dressing. Why does it look this way? The two liquids are layered in the pre-shaken bottle. Oil is less dense than vinegar and therefore floats on the top. The density of vegetable oil is .91 g/cm^3. The density of vinegar is 1.01 g/cm^3. You might want to have a bottle of salad dressing to show this separation.

Situation 3: An iron nail sinks to the bottom of a jar of water but a boat made with iron can float on top of water. How is this possible? The density of iron is 7.87 g/cm^3; therefore, iron nails will sink in water. Although the boat is constructed with iron, there are air compartments within the interior of the boat as well. A boat made of iron has an **average** mass to volume ratio less than 1.0 g/cm^3; therefore, it floats.

Teaching Tips

1. The specifications for the supertanker *Seawise Giant* are described in the introduction. Pictures of large cruise ships and tankers can be displayed to stimulate student discussion.

2. A clear tub or aquarium of water with a variety of objects sinking and floating can be used as a visual to trigger student thought.

Lesson Two: The Floating Grape
Student Investigation

Focus Questions
1. Under what conditions will a grape float?
2. Under what conditions will a grape sink?

Predictions
1. I think a grape will float if _____because _____.
2. I think a grape will sink if _____because_____.

SAFETY ALERT: NEVER taste any lab chemicals, even if you know what they are. Chemicals used in the lab are often reused and unsanitary for human consumption.

Materials per group: Safety goggles, tray to perform investigations, paper towels, plastic spoon, 3 clear plastic cups (5-8 oz), 1 cup of prepared salt water, 1 small cup of dry salt, 250 mL beaker of water, 250 mL beaker for waste, grape, electronic balance, 100 mL graduated cylinder

Procedure 1. Observe a Grape in Water
1. Place a piece of paper towel under a plastic cup.
2. Pour water into the cup until it is 3/4 full.
3. With the plastic spoon, transfer a single grape to the cup.
4. Observe what happens and record your findings in your science notebook.

Procedure 2. Observe a Grape in Salt Water
1. Place a piece of paper towel under a second plastic cup.
2. Pour salt water into the cup until it is ¾ full.
3. With a plastic spoon transfer the grape from the tap water cup to the cup with salt water.
4. Observe. Record your findings in your science notebook.

Procedure 3. A Hanging Grape
Consider your observations in the first two procedures.
Focus Question: How can you get the grape to "hang" in a solution cup? In other words, the grape will neither float nor sink, but hang suspended.
Prediction: Under what conditions will the grape "hang" in a liquid?

1. With your partner, design a plan to test your prediction. In your plan, describe the steps to get your grape to "hang" in solution. Your teacher will sign off on your plan before you proceed.
2. Try out your idea.
3. Did your idea work? If yes, go to step 5. If no, go to step 4.
4. Continue problem solving with the materials you have until you are successful. Record the changes you make. **Do not discard the hanging grape solution until you complete step 5.**

5. **Calculate the Mass of the Hanging Grape Solution:** Before discarding your successful hanging grape's solution, you need to calculate the mass of the solution. Place an empty graduated cylinder on the balance. Re-zero the balance and add 100 mL of your solution. Record the mass of the hanging grape's solution in your science notebook. You will need this measurement in the next investigation. *After recording the solution's mass, you can discard the solution.*

Claims and Evidence: Under what conditions can you claim a grape will float? Will sink? Will hang suspended in a solution?

New Understandings
1. Why do you think the grape sank in the tap water?
2. Why do you think the grape floated in the salt water?
3. Why do you think the grape just "hangs?"
4. Would you expect the same results with a sugar water solution and a grape? Explain.
5. Would you expect the same results with a different type of fruit – say an apple?

Reflections: What can you say about the scientific processes you used to get the grape to hang?

Teacher Notes: The Floating Grape
Lesson Two: Student Investigation

Rationale: This investigation introduces students to density by engaging them in a problem-solving situation. This lesson sets the stage for concept formation and serves as a before reading activity.

Materials & Preparation: Recipe for salt solution: 25 g of salt per 100 mL of water. Two to three liters of solution will be enough for a class of 30 students.
1. Distribute grapes after students have written their predictions.
2. Supply a 250 mL beaker or container for waste.
3. In procedure 3 students make a written plan to get the grape to hang before they actually do it. Sign off on their plan before they proceed. A small cup or container of 25 g of salt should be sufficient for this step. Extra salt should be available at the supply table.
4. Students will need to find the mass of 100 mL of their own prepared solution (the solution in which the grape "hangs") before they discard it (step 5 - procedure 3). The mass of their "hanging grape solution" will be added to a data chart in Lesson Four.

Teacher's Role: In this activity, you will facilitate the lab work by managing the supplies and their use, observing student work, and asking probing questions. As you monitor student work, continually assess student understanding as evidenced by their comments, questions, and notebook entries.

Background Information: Density is traditionally expressed as g/cm^3. However, students will frequently measure volume in milliliters (mL). Reinforce that 1 mL is equal to 1 cm^3. Grapes will sink in tap water and float in the prepared salt water. The successful hanging grape solution will have the same density as the grape.

Answer Key and Sample Responses

Claims and Evidence
Under what conditions can you claim a grape will float, sink, or hang? The grape will float in the prepared salt solution, sink in the tap water, and "hang" in a solution with a mixture of salt and water.

New Understandings
1. **Why do you think the grape sank in the tap water?** Students will most likely answer that the grape is heavier than the tap water and lighter than the salt water. A few students may use the term density at this point, but it is not necessary to define it until all students complete the reading that follows.
2. **Why do you think the grape floated in the salt water?** Students will most likely answer that the grape is lighter than the salt water. Few students will use the term density at this point.
3. **Why do you think the grape just "hangs?"** Student answers will vary.

Lesson Three: Defining Density
Student Reading

In Lesson Two you found that grapes float or sink depending on the liquid. What is it about the grape and the liquid that determines whether it will sink or float?

As you read, answer the following questions in your notebook.
1. What is density?
2. What two measurements do you need to determine density?
3. Why did your grape float in the salt water?
4. Why do objects float differently in the ocean as compared to fresh water?

Scientists use the term density when explaining whether something will sink or float. Density is a characteristic property of a substance. It is the ratio of the mass of an object compared to its volume. Mass is the amount of matter contained in an object and is commonly measured in grams (g) with a balance. Volume is the amount of space taken up by the matter and is measured in milliliters (mL) or cubic centimeters (cm^3). One mL equals one cm^3.

Density is expressed as a numerical value by using this formula
Density (D) equals mass (M) divided by volume (V).

Mass=10 g
Volume= 2 cm^3

Box A

Mass=16 g
Volume= 2 cm^3

Box B

The two boxes have the same volume, but they have different masses. The density of box A is 10g /2 cm^3 = 5 g/cm^3. The density of box B is 16 g/2 cm^3 = 8 g/cm^3.

We can use the term density to describe the grape's behavior.

- When the grape's density is **greater** than the density of the liquid, the grape **sinks.**
- When the grape's density is **less** than the density of the liquid, the grape **floats.**

Building a Semantic Map: Semantic maps are visual tools that show connections among ideas, events, and data.

To Do

1. Obtain a copy of the semantic map outline.
2. Using information from your investigations and the reading, fill in as much as you can. The book symbol on the map indicates you will find information from the readings. The magnifying glass symbol means this information is retrieved from your investigations.
3. As you complete more lessons, you will add to your map.

Semantic Map Outline

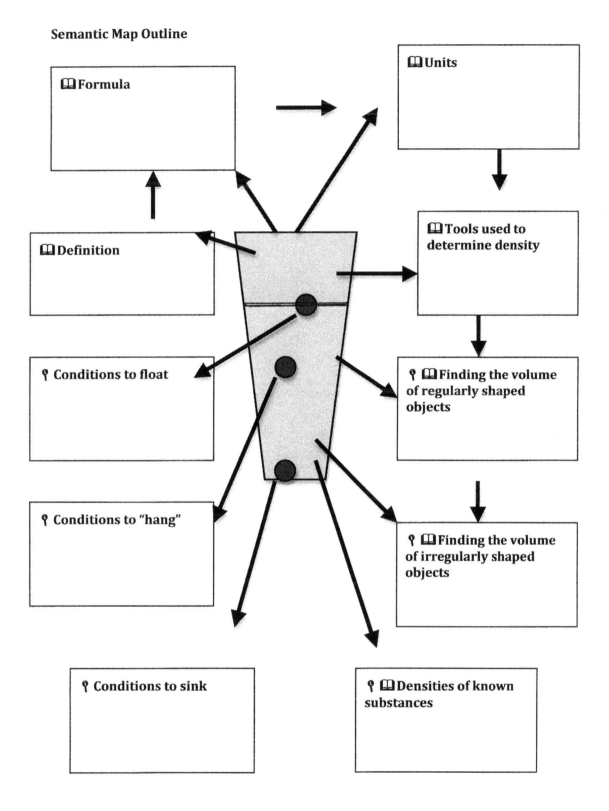

Teacher Notes: Defining Density
Lesson Three: Student Reading

Rationale: Students are provided with a definition of density, a visual model of density, and a mathematical formula they can use to calculate density.

Reading Strategies: Hands-On Investigation, QAR (Question-Answer Relationships) and Semantic Map. See Literacy Resource Section.
    Before Reading: Lesson Two: The Floating Grape
    During Reading Strategy: QAR (Question-Answer Relationships)
    After Reading Strategy: Semantic Map

Materials & Preparation: Student copies of the Reading: Defining Density with QAR Questions and Semantic Map Outline

QAR Teacher's Role: Link the previous lesson to the reading. Highlight the four QAR questions at the beginning of the reading and prompt students to answer the questions from their reading. Discuss the questions after the reading.

QAR Questions and Sample Responses
1. **What is density**? (Right There) Density is the ratio of the mass of an object compared to the volume it occupies.
2. **What two measurements do you need to determine density?** (Think and Search) mass and volume
3. **Why did your grape float in the salt water?** (Author and You) The grape was less dense than the salt water.
4. **Why do objects float differently in the ocean as compared to fresh water?** (On My Own) The ocean and fresh water have different densities.

Semantic Map Teacher's Role: Introduce the semantic map as an organizational tool. Guide students to the boxes that can be filled in at this point: definition, formula, units, conditions to float, conditions to sink. Point out *book* and *magnifying glass* symbols as clues to where information is presented. The *book* symbol on the map indicates you will find information from the readings. The *magnifying glass* symbol means this information is retrieved from investigations. Let students know that their semantic maps will be completed in a later lesson.

A key to the semantic map can be found in the Teacher Notes: Lesson Nine. The semantic map format can be differentiated for the student audience. Some students can design their own maps; others may need the template. If students design their own maps, list the types of information that will be required on the finished map: definition, formula, units, tools, known densities, etc.

Lesson Four: Calculating Density
Student Investigation

In Lesson Two you observed that a grape would sink in tap water and float in salt water. The density of the grape and the density of the liquid can be used to explain the sinking and floating. In this investigation you will calculate the actual density values of the tap water, the salt water, and the grape. Predicting whether a substance will float or sink in a liquid can be determined by comparing their density values.

Focus Questions
1. How is the density of a substance measured in a laboratory?
2. How do the calculated density values of tap water, salt water, and the hanging grape solution compare to the density of the grape?

Predictions:
1. I think the grape will have a density value (greater than/the same as/less than) the tap water because...
2. I think the grape will have a density value (greater than/the same as/less than) the salt water because...
3. I think the grape will have a density value (greater than/the same as/less than) the hanging grape solution because...

Part 1. Calculating the Densities of Tap Water, Salt Water, and Hanging Grape Solution
Materials: 100 mL graduated cylinder, 2 plastic cups, tap water, prepared salt water solution, plastic spoon, electronic balance, calculator, grapes

Observations: Copy the chart below in your science notebook. You will use it to record measurements and calculate density.

Part 1. Calculating Densities			
Liquid	mass (g)	volume (cm^3)	Density (g/cm^3) mass/volume
Tap water (100 mL)			
Prepared Salt Solution (100 mL)			
Hanging Grape Salt Solution (100 mL)	From Lesson Two		

Procedure 1. Calculating the Density of Tap Water
1. Place an empty graduated cylinder on the electronic balance.
2. Tare (re-zero) the balance.
3. Remove the graduated cylinder from the balance and add 100 mL of tap water.
4. Place the cylinder with the tap water on the tared balance. You will now have the mass of just the tap water. Record the tap water mass in grams in your chart.
5. Record the volume of the tap water in the chart. The graduated cylinder measurement markings are in milliliters. You will need to convert the milliliter (mL) measurement to cubic centimeters (cm^3) because density units are typically reported as grams per cubic centimeter. It is an easy conversion: one milliliter (mL) is equal to one cubic centimeter (cm^3). For example: 46 mL of water is equal to 46 cm^3 of water
6. Calculate the density of tap water using the formula for determining density: mass divided by volume. Do this calculation and record your answer on the data chart.

Procedure 2. Calculating the Density of Salt Water
1. Repeat steps 1- 4 in procedure 1 to find the mass of salt water in grams.
2. Record the volume of the salt water in cm^3.
3. Calculate the density of the salt water as you did in step 6 in procedure 1.

Procedure 3. Calculating the Density of Hanging Grape Water
1. In your science notebook, find the mass of your "hanging" grape salt water solution recorded in Lesson Two. Record the mass in the data chart.
2. Record the volume of the "hanging" grape salt water in the data chart.
3. Calculate the density of the "hanging" grape salt water as you did in step 6 in procedure 1.

Part 2. Student Reading: What About the Density of the Grape?

To calculate the density of a substance in the laboratory we need to know its mass and volume. A balance is used to determine mass, but finding the volume of a solid substance requires a different procedure. You can't "pour" a solid into a graduated cylinder and measure it's volume in mL.

The volume of **regularly** shaped solids is determined by using mathematical formulas. For example, the formula for finding the volume of a rectangular substance, like a brick, is determined by multiplying the length x width x height of that solid.

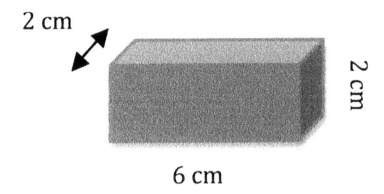

By multiplying the height (2 cm) times the width (2cm) times the length (6 cm) of this block, its volume is calculated to be 24 cm³.

Determining the volume of a grape can be a challenge. The grape is a solid object with an **irregular** shape. Have you ever sat in a bathtub of water and seen the water level rise? Greek mathematician Archimedes did. He realized that the level the water rose was equal to the volume of his body (an irregular shaped solid). Scientists use a similar procedure called the **water displacement method** to find the volume of irregularly shaped solid objects.

A solid substance can be added to a graduated cylinder with a known volume of water. The water will be displaced by the solid and rise. The difference between the original volume of water and the new volume (with the solid) is the volume of the solid.

A graduated cylinder contains 53 mL of water. When a marble is put into the cylinder, the water level rises to 60 mL. The difference between the beginning measurement of the water and the ending measurement with the marble will be the volume of the marble: 60 mL – 53 mL = 7 mL or 7 cm³. The mass of the marble is determined with a balance. Let's assume the marble has a mass of 17.5 g. You can now calculate the density of the marble. The density is 17.5 g / 7 cm³ or 2.5 g / cm³.

Part 3. Calculating the Density of the Grape
Now that you have calculated the densities of the different salt water solutions, it is time to calculate the density of a grape. Since grapes are irregularly shaped solids, you will use water displacement to find its volume.
Observations: Copy the chart below in your science notebook. Use it to record measurements and calculate density.

Part 3. Calculating the Density of a Grape					
	Mass (g)	Beginning volume of water in graduated cylinder (mL)	Volume in graduated cylinder after grape is added (mL)	Difference in volume (cm^3)	Density (g/cm^3) (mass/volume)
Grape		50 mL			

Procedure. Calculating the Grape's Density
1. Find the mass of the grape on the electronic balance. Record the mass in the data chart.
2. Put 50 mL of tap water into a 100 mL graduated cylinder.
3. Carefully lower the grape into the cylinder.
4. Record the new water level in the graduated cylinder on your chart.
5. Subtract the original 50 mL of water from the new water level. This value is the volume of the grape. Record the volume of the grape on your chart.
6. Determine the density of the grape using the density formula. Record on your chart.

Claims and Evidence
1. Write a statement comparing the calculated density values of the tap water, teacher prepared salt water solution, and "hanging" grape solution.
2. Write a statement comparing the density of the grape with its behavior in the teacher prepared salt water solution and the "hanging" grape solution.

New Understandings
1. Why did the grape "hang" in your team's solution?
2. What would happen to the density of a sample of salt water if you added tap water to it? Explain your reasoning.
3. Why wasn't the density of every lab group's "hanging" salt solution the same?
4. If the mass (g) of a substance is greater than it's volume (cm^3), will the substance sink or float in tap water? Explain your reasoning using the formula for density.

Reflections
1. What questions do you have about density?
2. What comments do you have about the scientific processes you used in this investigation?

Teacher Notes: Calculating Density
Lesson Four: Student Investigation

Rationale: In this investigation students learn how to use the formula (density equals mass divided by volume) to determine the densities of the substances used in Lesson Two. Using quantitative data, they will scientifically explain why a substance sinks and floats in a particular liquid. A short reading explaining how to determine the density of irregularly shaped solids is integrated into this lesson. Water displacement is used to determine the volume of the grape.

Part 1. Calculating the Densities of Tap Water, Salt Water, and Hanging Grape Solution
Part 3. Calculating the Density of the Grape

Materials and Preparation: Student copies of Lesson Four, 100 mL graduated cylinder, 2 plastic cups, tap water, prepared salt water solution (the same concentration used in Lesson Two – 25 g salt in 100mL water), plastic spoon, electronic balance, calculator

Teacher's role

In this activity, you will facilitate the lab work by managing the supplies and their use, observing student work, and asking probing questions. You will demonstrate the water displacement technique for determine the volume of an irregular solid. Model collecting data and calculating density. As you monitor student work, continually assess student understanding as evidenced by their comments, questions, and notebook entries.

1. Work through a few sample problems for determining density before students calculate their own density values.
2. Students may need instruction on how to put measurements into their calculators.
3. If necessary, demonstrate how to tare or re-zero an electronic balance.
4. Assist students in applying the displacement technique to determine the volume of a grape.

Part 2. Student Reading: What About the Density of the Grape?
Reading Strategy: Hands-on Investigation

Teacher's Role: After completing Part 1 of Lesson Four, facilitate a discussion about how to find the volume of a solid. Demonstrate the technique to determine the volume of an irregular solid by water displacement with an irregularly shaped object like a rock.

Background: See the Student Reading: What About the Density of the Grape?

Parts 1-3. Answer Keys and Sample Responses

Claims and Evidence: The density values below were accurately determined from experimental data; however, student measurements may differ slightly. The density of tap water should be less than the density of the prepared salt water. The "hanging" solution should have a density between the tap and prepared salt solution. If this is not the pattern, you may want them to go back and check their mathematics and/or measurements.

1. **Write a statement comparing the calculated density values of the tap water, teacher prepared salt water solution, and "hanging" grape water.** The density of tap water is 1.0 g/cm^3, the density of the prepared salt water is 1.12 g/cm^3, and the density of the "hanging" grape solution should be between 1.0 and 1.12 g/cm^3.

2. **Write a statement comparing the density of the grape with its behavior in the teacher prepared salt water solution and the "hanging" grape solution.** When the density of the grape was the same as the salt solution, it did not sink or float but was suspended in the solution. When the grape's density was greater than the density of the water, it sank. When the grape's density was less than the salt water, it floated.

New Understandings

1. **Why did the grape "hang" in your team's solution?** The density of the grape and the density of the "hanging solution" were the same.

2. **What would happen to the density of a sample of salt water if you added tap water to it? Explain your reasoning.** The density would decrease because the salt particles would be spread out in more volume. When the density of 100 mL of the solution is calculated, there will be fewer particles packed into the solution. You might introduce the term dilution here.

3. **Why wasn't the density of every lab group's "hanging" salt solution the same?** The equipment and measurement techniques most likely affected our calculations.

4. **If the mass of a substance is greater than it's volume, will the substance sink or float in tap water? Explain your reasoning using the formula for density.** Using the formula for density if the mass value is larger than the volume value, the density will always be greater than one. The density of water is 1.0 g/cm^3. Anything with a density greater than this will sink in water.

Teaching Tips

1. Make sure the graduated cylinders are large enough to submerge the grape (50 - 100mL).

2. **Optional Demonstration:** Immerse a can of diet pop and it's sugared counter part (i.e., Diet Coke™ vs. regular Coke™) in a clear tub of water. The diet drink will float; the sugared drink will sink. Have students suggest reasons why this is happening. The correct answer is that the sugared drink is denser – more mass per the same volume. The density of the diet drink is less than the density of water; thus, it floats.

Lesson Five: Does Density Change?
Student Investigation

Focus Questions
1. How does the amount of a substance affect its density?
2. How does the shape of a substance affect its density?

Predictions
1. I predict that 100 mL of water will have a density (greater than / the same as / less than) the density of 50 mL of water because _____.
2. I predict that a long piece of clay rope will have a density (greater than / the same as / less than) a shorter piece of clay rope because _____.

Materials per group: 100 mL graduated cylinder, electronic balance, modeling clay in one color, metric ruler, calculator.

Observations/Data
Copy the chart below in your science notebook. Record measurements and calculate the densities.

Calculating the Density of Water		
Mass (g)	**Volume (cm³)**	**Density (g/cm³) Mass/Volume**
	25 mL (cm³)	
	50 mL (cm³)	
	75 mL (cm³)	
	100 mL (cm³)	

Procedure 1. Calculating the Density of Water
1. Find the mass of an empty 100 mL graduated cylinder on the balance. *Re-zero* the balance. Add **25 mL** of water to the cylinder and find its mass. Record on the data chart.
2. Find the mass of an empty 100 mL graduated cylinder on the balance again. *Re-zero* the balance. Add **50 mL** of water to the cylinder and find its mass. Record on the data chart.
3. Find the mass of an empty 100 mL graduated cylinder on the balance again. *Re-zero* the balance. Add **75 mL** of water to the cylinder and find its mass. Record on the data chart.
4. Find the mass of an empty 100 mL graduated cylinder on the balance again. *Re-zero* the balance. Add **100 mL** of water to the cylinder and find its mass. Record on the data chart.
5. Calculate the density of all four volumes of water by dividing the water's mass by its volume.
6. Share your data with the class on the chart provided by your teacher.

Procedure 2. Calculating the Density of Modeling Clay

1. Copy the chart below in your science notebook.
2. Using a single color of modeling clay, make three different clay ropes of different lengths. Be sure each rope will easily fit inside the 100 mL graduated cylinder.
3. Measure the length, the mass, and the volume of each clay rope. Use the water displacement technique to measure the volume of each clay rope. Record.
4. Calculate the density of each rope and record in the data chart.
5. Now, let's use another method to determine the volume of a **rectangular shape of clay.** First, you will need to mold a rectangular box-shaped piece of clay. With a ruler measure its length, width, and height. Calculate the volume and mass of the clay, and then calculate its density. Record all your measurements in your science notebook. Hint: Volume of a rectangular solid equals length x width x height.

Calculating the Density of Modeling Clay Ropes				
Rope	Length (cm)	Mass (g)	Volume from displacement in a graduated cylinder (cm^3)	Density of clay rope mass divided by volume (g/cm^3)
1				
2				
3				

Claims and Evidence

1. How does the **volume** of water affect its density?
2. Does the length of the clay affect its density? What evidence leads you to this claim?
3. How does the density of the rectangular clay shape (calculated using the mathematical formula) compare to the density of the clay determined by displacement?

New Understandings

1. What is the density of water? Is this the same regardless of amount? Explain your reasoning.
2. What is the density of the modeling clay you tested?
3. Would you expect all colors of modeling clay to have the same density? Check with a group that had a different color than you had. Explain.
4. Would you expect all brands of clay to have the same density? Explain.
5. Will salt water always have the same density? Use evidence from the teacher prepared salt water solution and the "hanging" grape solution used in Lesson Four in your answer.
6. The density of water is 1.0 g/cm^3. Did your lab group get this exact density? What might have influenced your results

Reflections

1. What questions are still unanswered?
2. How might density values be useful?

Teacher Notes: Does Density Change?
Lesson Five: Student Investigation

Rationale: Students discover density is a characteristic property of matter regardless of amount or shape. Students apply the skills learned in the previous lessons to find mass and volume measurements, calculate density, and then compare their findings.

Materials and Preparation: Student copies of Lesson Five, 100 mL graduated cylinder, electronic balances, modeling clay (one color per group distributed in plastic bags), metric rulers, calculators.

1. Begin this investigation with three different containers of water such as a glass, a pitcher, and a jar. Ask students if the density of the water in the containers is the same or different. How could they test their idea?
2. For the student investigation provide 100 mL graduated cylinders that are calibrated in 1 mL increments.
3. Students will find the density of water for four different volumes. Measurements will not be identical, but they should be close.
4. Distribute the clay in a small sealable plastic bag to each lab group. Groups should have only one color of the clay. This eliminates color as a variable. Oil-based modeling clay is ideal for this investigation as it is reusable, won't dry out, harden or dissolve in water. It is recommended that you personally collect the clay at the end of the investigation.
5. Review measuring with a metric ruler. Students should record their answers in centimeters.
6. Combine group results and conduct a discussion on the need for multiple trials and averaging the data.

Teacher's role: In this activity, you will facilitate the lab work by managing the supplies and their use, observing student work, and asking probing questions. It is recommended that you review how to collect data and calculate density. Facilitate a discussion that compares individual group results with class results. As you monitor student work, continually assess student understanding as evidenced by their comments, questions, and notebook entries.

Background
Density is a characteristic property of matter measured at standard temperature and pressure. The density for common substances such as vinegar, oil, and ammonia can be found in reference materials. Density will change with temperature as indicated in the math integration activity in the next lesson.

Constructing a rectangular solid from the clay reinforces finding volume using mathematical formulas. The density of all clay shapes and ropes within a lab group should be the same. Discrepancies will occur because of measurement inconsistencies.

Answer Key and Sample Responses
Claims and Evidence
Density values might be slightly off due to limitations of measuring equipment and experimenter error.

1. **How does the volume of water affect its density?** Density does not change with volume. Student responses should be near 1.0 g/cm^3 for all volumes tested.

2. **Does the length of the clay affect its density? What evidence leads you to this claim?** No, all densities should be the same.

3. **How does the density of the rectangular clay shape (calculated using the mathematical formula) compare to the density of the clay determined by displacement?** Density values should be the same or similar to that of the clay ropes.

New Understandings

1. **What is the density of water? Is it the same regardless of amount? Explain your reasoning.** The density should be close to 1.0 g/cm^3. It should be the same regardless of amount because density is a characteristic property of matter.

2. **What is the density of the modeling clay you tested?** Will vary depending on the type of clay.

3. **Would you expect all colors of modeling clay to have the same density? Check with a group that had a different color than you had. Explain.** Since the color added to the clay is a dye and the dyes are different, the clays have different chemical compositions. Thus, their densities could be different. However, the quantity differences of the dyes may be so small as to not affect the class measurements. They most likely will be the same.

4. **Would you expect all brands of clay to have the same density? Explain.** Probably not as they have different ingredients or different amounts of similar ingredients.

5. **Will salt water always have the same density? Use evidence from the teacher prepared salt water solution and the "hanging" grape solution in Lesson Four in your answer.** No. Salt water can be of varying densities.

6. **The density of water is 1.0 g/cm^3. Did your lab group get this exact density? What might have influenced your results?** Precision of equipment, accuracy of measurements, impurities in water, etc.

Lesson Six: Making Sense of Density Data
Student Chart Reading, Graphing, and Data Interpretation

Scientists can access density values for substances from the periodic chart and from reference materials. Density is an identifying characteristic property of matter.

Chart Reading: Use the chart below to answer the questions that follow.

Densities of Common Metals	
Substance	Density g/cm^3
Iron	7.8
Aluminum	2.7
Lead	11.3
Gold	19.3
Silver	10.5
Copper	8.9
Nickel	8.8
Zinc	7.0

1. Which metal is the most dense?

2. Which metal is the least dense?

3. Will any of these substances float in water? Explain your thinking.

4. How could you use the information in the table to determine if something is gold or copper?

5. Using the reference table above, identify an unknown metal with the following measurements: mass = 133.5 g and volume = 15 cm^3.

6. The density of iron is 7.8 g/cm^3, and the density of water is 1.0 g/cm^3. Why do boats made of iron float?

Your lab group found that the density of water was near 1.0 g/cm³. Although the density of water is the same regardless of the amount, scientists know that the density of water changes with temperature.

Graphing the Data: In your science notebook, make a graph of the data below.

Water Temperature and Density	
Temperature °C	Density (g/cm³)
25	1.00
50	.98
75	.97
100	.97
125	.95
150	.92
175	.88
200	.86

Data Interpretation

1. What happens to the density of water as temperature increases?

2. What is the total change in density from 25°C to 125°C?

3. Between what temperatures is there the greatest change in density?

4. Which floats on top of the other: cold water or hot water? Explain your answer.

5. Predict the density of water at 250°C.

6. Predict the density of water at 110°C.

7. Based on the chart and your graph, what can you predict about the density of ice?

8. If salt water is heated, will its density change? Explain your thinking.

Teacher Notes: Making Sense of Density Data
Lesson Six: Student Chart Reading, Graphing, and Data Interpretation

Rationale: Students analyze density data using charts and graphs. The graphing exercise provides students with practice in graphing data and analyzing the results. The data interpretation questions are of increasing difficulty to challenge student thinking.

Materials and Preparation Student copies of Lesson Six: Making Sense of Density Data, graph paper and metric ruler.

Teacher's role: Engage student interest by reading the classic story of *Archimedes and the Golden Crown* described below before handing out the copies of the lesson. This story is an interesting way to introduce the chart reading activity. Facilitate the activity by observing student work, and asking probing questions. If necessary, model how to set-up a graph. As you monitor student work, continually assess student understanding as evidenced by their comments, questions, and notebook entries.

Archimedes and the Golden Crown
Archimedes was brought in by King Hiero (3rd century B.C.) to find out if the jeweler, who was commissioned to make him a golden crown, had tricked him. Archimedes placed a solid gold bar the same mass as the crown into water and measured the water overflow. He then placed the golden crown in water and measured the water overflow. The amounts of displaced water (the overflow) were not the same. The crown was a fake.

At the beginning of Lesson Six, is a chart providing the densities of a variety of metals. Students can be asked to suggest which metal might have been substituted for gold.

Background
The graphing exercise reveals that the density of *liquid water* does change with a change in temperature. The explanation for this is found in the kinetic molecular theory. When water is cooled, molecular motion slows and the water molecules move closer together increasing the density. When water is heated, the water molecules gain energy and move farther apart (and move faster), lowering the density; thus, there are more molecules in a given volume of cool water vs. hot water. In general, the solid phase of a substance has a greater density than its liquid phase. The exception is the density of ice ($.92 \text{ g/cm}^3$) and a few other substances like bismuth. *Do not share the density value of ice at this time as it is explored in the next lesson.*

Answer Key and Sample Responses
Chart Reading and Data Interpretation
1. **Which substance is the most dense?** gold
2. **Which substance is the least dense?** aluminum

3. **Will any of these substances float in water? Explain your thinking.** As pure substances, they all have a density greater than the density of water (1.0 g/cm^3). None of the substances will float in water.

4. **How could you use the information in the table to determine if something is gold or copper?** Find the density of the unknown substance and compare it to the chart for gold and copper.

5. **Using the chart, identify an unknown substance with the following measurements: mass = 133.5 g and volume = 15 cm^3.** 8.9 g/cm^3 copper

6. **The density of iron (steel) is 7.8 g/cm^3 and the density of pure water is 1.0 g/cm^3. Why do boats made from iron float? Hint: revisit the formula for density.** Students may need help with this one. Although the density of iron is greater than the density of water, a boat is contains many air compartments. The density of air is less than the density of water. The average density of the entire structure (iron, wood, air, etc.) must be less than 1.0g/cm^3 for the boat to float. Surface area and buoyancy also play a role, but are not addressed here.

Graphing the Data

Density of Water vs. Temperature

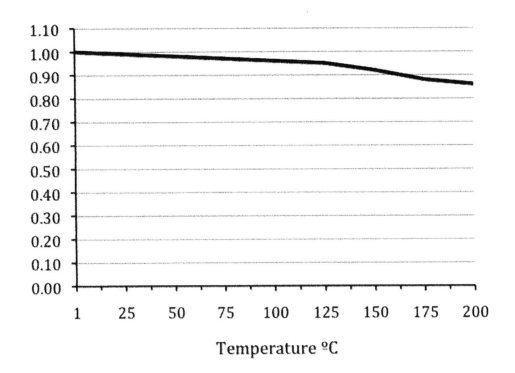

Density
g/cm^3

Temperature ºC

Data Interpretation
1. **What happens to the density of water as temperature increases?** Density decreases.
2. **What is the total change in density from 25°C to 125°C?** 0.05 g/cm^3
3. **Between what temperatures is there the greatest change in density?** 150 -175°C

4. **Which floats on top – cold water or hot water? Explain your answer.** Hot water is less dense than cold water; therefore, it will float on top.

5. **Predict the density of water at 250 °C.** Less than .86 g/cm³

6. **Predict the density of water at 110 °C.** Between .95 g/cm³ and .97 g/cm³

7. **Based on the chart and your graph, what can you predict about the density of ice?** Students may predict that the density of ice should be greater than 1.0 g / cm³ based on the data pattern. However, students should also know that ice floats. Thus, they may say that ice must have a density less than 1.0 g/cm³. The density of ice will be explored in the next lesson.

8. **If salt water is heated, will its density change? Explain your thinking.** Since salt water is partly water, it should follow a similar pattern. Warm salt water should be less dense than cold salt water.

Teaching Tips

1. Detailed versions of Archimedes story can be found on the Internet.

2. **Optional Demonstration:** To introduce the graphing exercise, add colored hot water to a clear container of ice-cold water. Slowly pour the hot water down the side of the container of cold water. The hot water will float on top. The greater temperature difference between the hot and cold water, the better. Practice beforehand to get optimum results.

3. Some students may need assistance in setting up the graph. Students should be encouraged to plot the points and then connect them with a line.

Lesson Seven: Issues with Ice
Student Investigation

In an earlier lesson, you found the density of water to be the same (at a given temperature) regardless of the amount of water. You also learned that density changes with water temperature. Ice, the solid form of water floats. What is special about the density of ice?

Focus Questions: What happens when you put ice cubes in liquids with different densities?

Materials (2-3 students per group): Clear tall plastic cup or cylinder with water, clear tall plastic cup or cylinder with vegetable oil, 2 colored ice cubes

Procedure
1. Gently lower a colored ice cube into the cup of water and the other colored ice cube into the vegetable oil cup.
2. Observe for 15 – 30 minutes.

Observations: Describe what happens in your science notebook using labeled drawings.

Claims and Evidence
1. What can you claim about the density of ice? State your evidence.
2. What can you claim about the density of oil? State your evidence.
3. What can you claim about the density of liquid water in relation to the density of oil?

New Understandings
1. Given ice, vegetable oil, and tap water, which is denser? How do you know?
2. Given ice, vegetable oil, and tap water, which is the least dense? How do you know?

Teacher Notes: Issues With Ice
Lesson Seven: Student Investigation

Rationale: This investigation allows students to continue to construct meaning through additional experiences. Here, students observe and explain the floating behavior of ice as compared to the sinking behavior of the water from its melt. In the graphing lesson, it was found that the density of water decreases with temperature. Note: Ice is an exception.

Materials and Preparation: Student copies of Lesson Seven: Issues with Ice, 2 clear plastic cups or glasses, vegetable oil, water, and colored ice cubes. Note: Make colored ice cubes a day before by adding food coloring to the water. Blue cubes are easily visible in the yellow oil. Tall, clear containers or glasses create a dramatic visual as the ice melts and sinks.

Teacher's role: Encourage students to record descriptive observations and to use diagrams. Model this method of data collection. Students should use the term density to describe why ice floats and why the melt water sinks. However, they may not know why the density of ice is less than the density of water. Let them share their ideas before you reveal the explanation. If time permits, allow students to explore some more (i.e., ice melting in another liquid such as vinegar).

Background: The oil provides a unique visual because water is insoluble in the oil. The density of ice is 0.92 g/cm^3; water is 1.00 g/cm^3. Vegetable oil has a density between 0.91 – 0.93 g/cm^3. Ice floats on the oil, the water melt sinks. Ice has a lower density than water because of its unique crystalline structure and strong *inter*-molecular forces due the polarity of water molecules and hydrogen bonding. This information is for teachers only. The explanation for polarity of molecules and hydrogen bonding is beyond the middle level curriculum.

Answer Key and Sample Responses
Claims and Evidence
1. **What can you claim about the density of ice? State your evidence.** Ice floats on top of the vegetable oil so it is less dense than the oil.
2. **What can you claim about the density of oil? State your evidence.** The oil is denser than the ice but is less dense than the liquid water. The liquid water drops below the oil layer.
3. **What can you claim about the density of liquid water in relation to the density of oil?** Liquid water is denser than the oil because it sinks to the bottom.

New Understandings
1. **Given ice, vegetable oil, and tap water, which is denser? How do you know?** The tap water is denser because both ice and oil float on top of it.
2. **Given ice, vegetable oil, and tap water, which is the least dense? How do you know?** Ice is the least dense because it floats on top of oil and on top of tap water.

Lesson Eight: Density to the Rescue
Student Investigation

You have investigated several concepts about density.
- Floating or sinking in a liquid medium is dependant upon the densities of the substances.
- Density can be calculated by dividing the mass of the substance by its volume.
- Density is a characteristic property of a substance and does not change with amount or shape of that substance.
- The density of water is 1.0 g/cm^3 under standard conditions.

Introduction

Thousands of oil and chemical spills occur each year around the world as a result of accidents or natural disasters. Plants and animals living in or near water habitats can be harmed when exposed to these spills. Water supplies for communities can be tainted. Cleanup of spills often requires costly and lengthy procedures. The Gulf of Mexico Oil Spill in April 2010 is a stark example of a devastating environmental catastrophe.

Focus Question: How can the property of density be useful in identifying and cleaning up liquid spills in the ocean or other bodies of water?

The Problem: Your team will be given three potential water pollutants. Your job is to create a density column of the pollutants and then use this information to design potential clean up strategies for each of the pollutants. *Common safe chemicals are used in this investigation. These chemicals represent real world pollutants like industrial wastes and crude oil.*

Materials Available (per group): Electronic balance, clear cylinder or test tube, graduated cylinder, empty clear cups, paper towels, soap, calculator, four plastic cups with the following liquids: 50 mL of water (blue), 50 mL each: pollutant A (clear), pollutant B (red) covered with plastic wrap, pollutant C (yellow), 4 pipettes or eyedroppers labeled for each liquid (A, B, C and water).

SAFETY ALERT: Pollutant B should be kept covered until you are ready to use it. This pollutant will evaporate easily and has a distinct odor. Do not inhale the vapors when transferring the liquid.

Procedure 1

Determine the density of each of the pollutants. Calculate the density for a volume of 50 mL. Record these values in your notebook.

Predictions: Using the density values calculated in procedure 1, predict how these three liquid pollutants, if spilled individually, would layer in a lake (fresh water).
1. When pollutant A is spilled in water, it will <u>float/sink</u> because _____.
2. When pollutant B is spilled in water, it will <u>float/sink</u> because_____.
3. When pollutant C is spilled in water, it will <u>float/sink</u> because_____.

Procedure 2

Test your predictions. Using a pipette, combine equal amounts of water and the pollutant you are testing in each prediction in a cylinder or test tube (for example, pollutant A with water). Record your observations in your notebook. Discard the mixture and test the other combinations one at a time. You may need to clean out your container with soap and water between tests.

More Predictions: How will these pollutants layer in a lake if they are all three spilled at the same time? Consider the density measurements you made in Procedure 1 as well as the experimental data you collected in Procedure 2. Remember to include water as one of the layers.

I predict that _____ will be the top liquid layer because_____.
I predict that _____will be the second to the top liquid layer because_____.
I predict that _____will be the third to the top liquid layer because_____.
I predict that _____will be the bottom liquid layer because_____.

Procedure 3

Design a plan to test your predictions and make a density column. Have your teacher approve your plan before you proceed. Use the materials provided and layer the liquids. Revise the plan until you have success layering the four liquids by their different densities. Draw and label the final layered liquid column in your notebook. Use appropriate colors to indicate the liquid layers.

Potential Clean up Strategies
1. Which pollutant would be easiest to clean up? Explain.
2. Which pollutant would be most difficult to clean up? Explain.
3. Describe a strategy that could be used to clean up one of the pollutants.

New Understandings
1. How would clean up be different if the spill occurred in the ocean (density of salt water is 1.02 g/cm^3) instead of fresh water (1.00 g/cm^3)?
2. A tanker truck spilled liquid into a river running along the highway. The liquid could be seen floating on the surface. What can you say about the density of this substance? How could you use the density of this pollutant to identify it?
3. Does the temperature of a lake affect the layering of the liquids? Explain.
4. How can the property of density be useful in identifying and cleaning up liquid spills in the ocean or other bodies of water?

Teacher Notes: Density to the Rescue
Lesson Eight–Student Investigation

Rationale: In this activity students apply their understanding of density to solve a real world problem. Students predict and test their ideas with a team-designed plan.

Materials and Preparation: Student copies Lesson Eight: Density to the Rescue, Electronic balance, clear cylinders or test tubes, graduated cylinder, clear cups, paper towels, soap, calculator, transfer pipettes, 50 mL of water with blue food coloring added, 50 mL each: pollutant A (clear) = corn syrup, pollutant B (red) = rubbing alcohol (covered with plastic wrap), pollutant C (yellow) = vegetable oil.

- **Prepare lab trays for student groups.** The four liquids used in this activity can be pre-measured (50 mL) and put into labeled plastic cups. Label the corresponding pipettes as well. Pollutant B, rubbing alcohol, should be covered with plastic wrap as the fumes may be irritating. Remind students to keep it covered. The classroom should be well ventilated for this activity.
- **A and B are actually soluble in water.** The layers exist temporarily. Over time, the layers will mix and cannot be separated by physical means involving density. The experiment must be completed before this occurs. Optimum layering will occur if students add the liquids in the correct order from most dense to least dense.
- **Provide clear cylinders**, such as a graduated cylinder or test tubes, for students to make density columns.
- **Supply liquid soap, hot water and paper towels for clean up.** The oil will create a greasy residue and the corn syrup is sticky. If available, supply student groups with more than one test tube or cylinder.
- **Liquids are colored with food coloring so that the layers are more readily visible.** Food coloring will not mix with the vegetable oil (yellow layer).
- **Revisit how to calculate density.** You may need to demonstrate this procedure again.
- **Review student plans before they begin creating their density column.** Model how to carefully add one liquid into another so that layering will occur. Let the liquid that is being introduced slowly flow down the side of the receiving container. Depending on the diameter of the cylinder used, it is recommended that the layers formed be uniform in height and easy to distinguish (minimum of 2 cm).

Teacher's Role: Engage student interest through the recommended scenario and discussion. Facilitate the lab work by managing the supplies and their use, observing student work, and asking probing questions. Reinforce safe lab procedures.

Begin this investigation by setting the scene. If there is a local water pollution issue, discuss this with your students. You may access visuals from the Internet as well. Ask your students how science can help solve real world issues. Lead the pre-investigation discussion with questions like: Can we use density to identify water pollutants? How? Could an understanding of density help in cleaning up pollution? Tell me more…

Background: Many bodies of water in the world have some level of pollution from chemicals and industrial waste. This real world issue is explored in this investigation from a density perspective. Real world pollutants are not safe to use in the classroom; thus, safe, liquids simulating these pollutants were selected. The density values of the three pollutants are: pollutant A, corn syrup = 1.33 g/cm^3, pollutant B, rubbing alcohol = 0.78g/cm^3 and pollutant C, vegetable oil = 0.91g/cm^3. Several industrial acids, such as sulfuric and nitric acid have density values greater than 1.3g/cm^3; therefore, sink in water. Gasoline has a density value in the range of 0.71–0.77 g/cm^3 and would float on water. The correct layering of the pollutants in the density column from the bottom up is corn syrup, water, oil, and rubbing alcohol (clear, blue, yellow, red).

Potential Clean up Strategies

1. **Which pollutant(s) would be easiest to clean up? Explain.** Pollutant B would be easiest because it floats on the top of the density column.
2. **Which pollutant would be the most difficult to clean up? Explain.** Pollutant A would be the most difficult because it is on the bottom of the density column. It may mix with all the other pollutants as it is pulled to the surface.
3. **Describe a strategy that could be used to clean up one of the pollutants.** Answers will vary. Answers may include removing pollutants by skimming, suction, filtering, absorption.

New Understandings

1. **How would clean up be different if the spill occurred in the ocean (density of salt water is 1.02g/cm^3) instead of fresh water (1.00g/cm^3)?** The oil layer is more distinct in salt water because there is a greater difference between the densities. This might make identification and clean up somewhat easier.
2. **A tanker truck spilled liquid into a river running along the highway. The liquid could be seen floating on the surface. What can you say about the density of this substance? How could you use the density of this pollutant to identify it?** The density of the liquid spilled is less than 1.0 g/cm^3 because it is floating on top of the water. If the density of the substance can be determined, it can be compared with the densities of known liquids.
3. **Does the temperature of a lake affect the layering of the liquids? Explain.** It might. The density of water increases as the water temperature decreases. It might be easier to separate an oil spill on a cold lake as the density difference between the oil and water would be greater than on a warm lake.
4. **How can the property of density be useful in identifying and cleaning up liquid spills in the ocean or other bodies of water?** When pollutants are analyzed for density, they can be matched to known density values – density is a characteristic property of matter. If a pollutant is less dense than water, it will float on top. Clean up procedures could entail skimming off the "floating" layer.

Teaching Tip: Lesson Extension. Provide students with a clear, plastic rectangular container of water with oil on top. Vegetable oil can be dyed with candle-making dyes (found in craft stores) to produce darker oil. Students design a plan to clean up the oil spill. Each lab group is given a variety of materials: string, cotton balls, liquid soap, feathers, suction devices (pipettes). Challenge student groups to produce the cleanest lake. There are several video clips of the Exxon Valdez Oil Spill (1989) and the Gulf of Mexico Oil Spill (2010) clean up efforts on the Internet. This would be a dramatic way to supplement the investigation.

Teacher Notes: Revisiting the Density Semantic Map
Lesson Nine: Student Review

Rationale: Students complete the semantic map started earlier in Lesson Three: Defining Density. They review the major ideas and concepts. The review prepares students for the post-assessment that follows.

Strategy: Semantic Mapping. See Literacy Resource Section.

Materials and preparation: Student copies of the semantic map outline introduced in Lesson Three: Defining Density.

Teacher's Role: Facilitate a discussion with your students on what they have learned during the entire Density Lesson Set. Have students follow the directions below to complete their maps.

Directions for Students:
1. Complete the map by adding information gained from Lessons four through eight.
2. Make additional boxes and connecting lines on the map, if needed. Add color and graphics.
3. Pair-share completed semantic maps.

Background: See example of a completed semantic map on the following page.

Teaching Tips
1. Stress the use of color and graphics.

2. Display the maps in your classroom.

Semantic Map Teacher Key

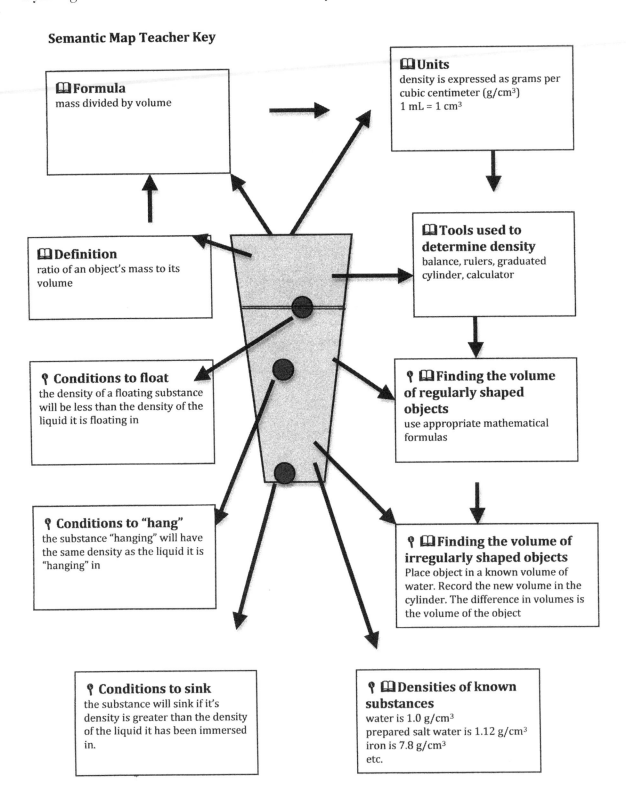

📖 **Formula**
mass divided by volume

📖 **Units**
density is expressed as grams per cubic centimeter (g/cm^3)
1 mL = 1 cm^3

📖 **Definition**
ratio of an object's mass to its volume

📖 **Tools used to determine density**
balance, rulers, graduated cylinder, calculator

♀ **Conditions to float**
the density of a floating substance will be less than the density of the liquid it is floating in

♀ 📖 **Finding the volume of regularly shaped objects**
use appropriate mathematical formulas

♀ **Conditions to "hang"**
the substance "hanging" will have the same density as the liquid it is "hanging" in

♀ 📖 **Finding the volume of irregularly shaped objects**
Place object in a known volume of water. Record the new volume in the cylinder. The difference in volumes is the volume of the object

♀ **Conditions to sink**
the substance will sink if it's density is greater than the density of the liquid it has been immersed in.

♀ 📖 **Densities of known substances**
water is 1.0 g/cm^3
prepared salt water is 1.12 g/cm^3
iron is 7.8 g/cm^3
etc.

Lesson Ten: Density Post-Assessment
Student Post-Assessment

Part 1. Multiple Choice: Circle the letter of the best response. If you wish to explain why you chose that answer, write your response by the question.

1. What is the best explanation for a layer of oil floating on top of water?
   a. The oil is denser than the water.
   b. The water is denser than the oil.
   c. The water is the same density as the oil.
   d. The oil is a different color than the water.

2. What is the best method for finding the **volume** of a medium sized rock?
   a. Use an electronic balance and record its mass.
   b. Measure its height and width. Multiply these two values.
   c. Grind the rock down and put "sand" into a graduated cylinder.
   d. Use the water displacement method.

3. The density of silver is 10.5 g/cm³. What will happen when a piece of silver is added to a beaker of water?
   a. The silver will float because its density is less than the density of water.
   b. The silver will "hang" because it has the same density as water.
   c. The silver will sink because it is denser than water.
   d. A small amount of silver will float in a large amount of water.

4. An estuary is an area where the fresh water of a river mixes with the salt water of the ocean. Which statement best describes the density of the water found in the estuary? The density of the estuary water is
   a. greater than the density of the river water.
   b. less than the density of the river water.
   c. the same as the density of the sea water.
   d. greater than the density of the sea water.

5. The amount of water vapor in the air is known as humidity. How does humid air compare to dry air in terms of density?
   a. Humid air is less dense than dry air.
   b. Humid air is the same density as dry air.
   c. Humid air is denser than dry air.
   d. Humid air does not have a density value.

6. Helium balloons float upward in air. How could you best explain the behavior of a helium balloon in terms of density?

    a. Helium is denser than air.
    b. Helium has the same density as air.
    c. Helium is a gas.
    d. Helium is less dense than air.

7. Thick syrup contains a large amount of sugar dissolved in water. Where would you find syrup in a density column with oil and water? The syrup will be
    a. at the top of the column.
    b. at the bottom of the column.
    c. mixed with the oil.
    d. in the middle of the column.

8. What is the density of a substance that has a mass of 10.0 g and a volume of 2 cm³?
    a. 0.5 g/cm^3
    b. 5.0 g/cm^3
    c. 20.0 g/cm^3
    d. 0.2 g/cm^3

9. Which of the following statements is *true* about density?
    a. A small piece of gold and a large piece of gold will have different densities.
    b. If you increase the volume of water, the density of the water will increase.
    c. An iron nail will have the same density as an aluminum nail.
    d. A silver coin will have the same density as a silver spoon.

10. A student wanted to find out if a ring is made of gold. How might the student determine if it is gold? Compare the ring's
    a. mass to the mass of a piece of gold.
    b. color to a piece of gold.
    c. volume to a known volume of gold.
    d. density to the density of gold.

Part 2. Agree or Disagree

Read each of the following statements carefully and determine if you (A) agree or (D) disagree. If you disagree with the statement, tell why you disagree below the statement.

1. _____ A grape with a density of .8 g /cm³ will float in a glass of water.

2. _____ The mass of a cube of aluminum is 5.4 g. Its volume is 2 cm³.
   The density of the aluminum is 10.8 g /cm³.

3. _____ Salt water is denser than tap water.

4. _____ A nail is dropped into a graduated cylinder that has 50 mL of water in it.
   The water rises to 53 mL. The volume of the nail is 3 cm³.

5. _____ An ice cube has the same density as water in a beaker.

Teacher Notes: Density
Student Post-Assessment

Part 1. Multiple Choice: Best Responses

1 = B, 2 = D, 3 = C, 4 = A, 5 = C, 6 = D, 7 = B, 8 = B, 9 = D, 10 = D

Part 2. Agree or Disagree

1. **Agree. A grape with a density of .8 g/cm^3 will float in a glass of water.**
2. **Disagree. The mass of a cube of aluminum is 5.4 g. Its volume is calculated at 2 cm^3. The density of the aluminum is 10.8 g/cm^3.** The density of the aluminum is 2.7 g/cm^3.
3. **Agree. Salt water is denser than tap water.**
4. **Agree. A nail is dropped into a graduated cylinder that has 50 mL of water in it. The water rises to 53 mL. The volume of the nail is 3 cm^3.** *Note: the conversion of mL to cm^3 has been reinforced throughout the lesson set because chemists report density in g/cm^3.*
5. **Disagree. An ice cube has the same density as water in a beaker.** An ice cube floats on top of water and therefore is less dense

Evaluation Suggestion

   Advanced: 13 -15
   Proficient: 10-12
   Partially Proficient: 6-9
   Unsatisfactory: 0-5

Lesson Set Three:
Attractive Forces

OVERVIEW

<u>Lesson Set Concepts</u>

3.1 Cohesive forces hold atoms and molecules of the same substance together.

3.2 Cohesive forces are strongest in solids and weakest in gases.

3.3 Unbalanced cohesive forces in liquids result in surface tension.

3.4 Surface tension is a characteristic property of liquids.

3.5 Surface tension can be calculated mathematically and used to compare liquids.

3.6 Surfactants are substances that reduce surface tension.

3.7 Water has a very strong surface tension.

3.8 Adhesive forces hold atoms and molecules of different substances together.

3.9 Adhesive forces between different substances allow substances to dissolve into each other. Sugar and water are examples of two substances with strong adhesive forces.

3.10 When adhesive forces between different substances are weak, they do not mix. Oil and water are examples of two substances with weak adhesive forces.

Key Vocabulary: cohesion, adhesion, surface tension, balanced force, unbalanced force, and attractive force

Lab Equipment and Supplies: Attractive Forces		
General Science Equipment & Supplies	**Grocery Store: Paper, Plastics, Specialty**	**Grocery/Box Stores/Pharmacy: Chemical Supplies**
Per Student: Safety Goggles Metric rulers Petri dishes	Tray for group materials Beakers, 250 mL Beakers, 15 mL Eyedroppers Cups, 1 oz. (catsup size) Toothpicks Paper towels Spoons, plastic Cups, plastic, 4 oz. Paper clips, Size #2 Sewing needle Small plastic/wood objects Wax paper Plastic wrap Aluminum foil Aluminum pie plates Optional items: leaf, fabric, coin, sand	Dishwashing liquid Pepper shaker, picnic size Milk, whole, 2%, 1% and skim Food coloring in dropper bottles

Lesson Set Guide: Attractive Forces

Lesson	Title	What Students Do
1 Intro Assess	What Would Happen If?	Students use their previous experiences to predict how water will behave in three different situations.
2 Invest	How Many Paper Clips?	Students investigate water's surface tension by comparing the number of paper clips that can be added to a cup filled with water to a cup filled with a dilute liquid dishwashing solution.
3 Read	Water's Amazing Invisible Elastic Skin	Students read about water's surface tension and summarize their reading.
4 Invest	Open Inquiry: Sinkers and Floaters	Students investigate the characteristics of objects that are denser than water but appear to float due to the strength of water's surface tension.
5 Literacy	Concept Map: Surface Tension	Students complete a concept map on surface tension.
6 Invest	Observing Water Drops	Students place drops of water on various surfaces. They compare the shape and height of the drops and relate their observations to the strength of attractive forces between the surfaces and the water drops.
7 Read	May the Forces be With You	Students read about cohesive and adhesive attractive forces and develop a main idea map.
8 Invest	A Detergent's Peppery Power	Students observe the effect of dishwashing liquid on water's surface tension in preparation for lesson nine.
9 Invest	Open Inquiry: Breaking the Surface	Students investigate the effect of a surfactant on different milk solutions: whole milk vs. reduced fat milks. They use the concepts of cohesion and adhesion to explain the interplay of the attractive forces between milk, detergent, and food coloring.
10 Read	From Milky Swirls to Washing Clothes	Student use an anticipation guide to read about adhesive forces and surface tension. They learn why milk swirls when food coloring and detergent are added and how detergent cleans clothes.
11 Chart Reading	Surface Tension Measurements	Students read and analyze quantitative data related surface tension. They compare and contrast the surface tensions of different liquids and learn how temperature affects the surface tension of water.
12 Review	Cohesion, Adhesion, or Both?	Students apply their learning by identifying whether an observation is an example of cohesive or adhesive attractive forces.
13 Assess	Attractive Forces Review	Students evaluate their learning with short answer response items and multiple-choice questions.

		Before and During the Lesson		End of Lesson
	Assessment Guide: Attractive Forces			
	Title	**Uncovers Student Ideas**	**Checks for new understandings**	**Evaluates learning**
1 Intro Assess	What Would Happen If?	Predictions 1, 2, 3		Pre-assessment repeated later in post-assessment.
2 Invest	How Many Paper Clips?	Prediction	NU: 1-6	C&E: 1-8
3 Read	Water's Amazing Invisible Elastic Skin	QARs	QARs	QARs Summary Paragraph
4 Invest	Explore Some More: Sinkers and Floaters	Focus Question Prediction Design	New Understandings	Claims & Evidence
5 Literacy	Concept Map: Surface Tension	Concept Map	Concept Map	Concept Map
6 Invest	Observing Water Drops	Predictions	NU: 2, 3	C&E: 1, 2 NU: 1
7 Read	May the Forces Be With You	Main Idea Map	Main Idea Map	Main Idea Map
8 Invest	A Detergent's Peppery Power	Prediction	NU: 1, 2	C&E
9 Invest	Explore Some More: Breaking the Surface	Prediction NU: 2	NU: 1	C&E
10 Read	From Milky Swirls to Washing Clothes	Anticipation Guide Parts I & II	Anticipation Guide Part III & IV	Anticipation Guide Part V
11 Chart Reading	Surface Tension Measurements	Q: 7	Q 1-7	Q 1-6
12 Review	Cohesion, Adhesion, or Both?	Q: 1-10	Q: 1-10	Q: 1-10
13 Assess	Student Post-Assessment			Parts 1, 2, 3

Lesson One: What Would Happen If?
Student Introduction and Pre-Assessment

Scientists are like detectives. Scientists use observation and questioning skills to explain real world events. In this set of lessons, you will become a "water" detective. Your job will be to observe and ask questions in order to explain the "unique" behavior of water under different conditions.

In earlier lessons you observed what happens when water is mixed with other liquids. Sometimes liquids, like vinegar, dissolve in water, and sometimes liquids, like oil, don't dissolve. As you observe water's behavior under different conditions, your focus question is:

What are the properties of water that allow it to behave the way it does?

Before you begin to investigate, let's use what you already know to make some predictions. In your notebook, describe in words or with drawings how you think water will act in the following situations.

Situation 1: A small cup of water is filled to almost, but not quite, overflowing. A paper clip is carefully dropped into the cup. The water does not overflow.
Prediction 1: How many paper clips do you think can be dropped into the cup before it overflows? Explain your thinking.

Situation 2: A student carefully places drops of water on two different surfaces: wax paper and plastic wrap.
Prediction 2: Will the water drops look the same or different on these surfaces? Explain your thinking.

Situation 3: Pepper is sprinkled on the surface of a small bowl of water. Next, a drop of dishwashing liquid is added to the water.
Prediction 3: What do you think will happen to the pepper when the dishwashing liquid is added? Explain your thinking.

Teacher Notes: What Would Happen If?
Lesson One: Student Introduction & Pre-Assessment

Rationale: This lesson both states the focus question for the lesson set as well as pre-assesses prior learning and experience. Before beginning a new lesson set, it is important to uncover student ideas.

Materials: None required for students.

Introductory Demonstration – Before Pre-Assessment
Materials: 4 clear punch cups: 2 for water, 1 for vinegar, and 1 for vegetable oil; blue food coloring, red food coloring.

Do this short demonstration before students make their predictions.
1. Fill two punch cups about ¼ full with water. Add blue food coloring to the water and stir.
2. To the third cup, add vinegar and add a drop of red food coloring to the vinegar and stir.
3. To the fourth cup, add some vegetable oil.
4. Before adding some red vinegar to one cup of blue water, ask the class what they think will happen. From previous lessons, students may remember than water and vinegar mix. The red vinegar and blue water will turn purple as the liquids mix.
5. Before adding the vegetable oil to the other cup of blue water, ask the class what they think will happen. From previous lessons, students may remember that water and oil will not mix. The oil will float on top of the water.
6. Finally, introduce the What Would Happen If lesson and have the students make their situation predictions in their science notebooks.

Teacher's Role: The teacher's role is to introduce the lesson and facilitate discussion as students make their predictions. The teacher should read student predictions to pre-assess student understanding.

Background Information: *Inter*-molecular attractive forces are introduced in this lesson set. *Intra*-molecular forces (chemical bonding) are not included in the lesson set. A common student misconception is that everything is held together by chemical bonds. By introducing *inter*-molecular attractive forces between atoms/molecules separately, there is a higher probability that students will be able to differentiate between physical and chemical changes.

Each situation described in lesson one is investigated in a later lesson. More specific explanations are provided in the later lessons.

Teaching Tips

- Have students write their individual predictions in their science notebook. When everyone is finished, conduct a brief class discussion to uncover ideas. Don't acknowledge correct or incorrect predictions at this time. Let students find out for themselves when they do the investigations to follow.

- As you read the student predictions, record their ideas for future reference. Also, record any "interesting" explanations.

Lesson Two: How Many Paper Clips?
Student Investigation

Focus Question: How many paper clips can be added to a FULL cup of water before the water overflows?

Prediction: I think _____ paper clips can be added to the cup of water before it overflows.

Materials per Group: Safety goggles, tray to perform investigations, paper towels, eyedroppers, toothpicks, water: 250 ml beaker or container of similar size, dishwashing liquid: 1 oz. "catsup size" cup, plastic spoon to measure and add dishwashing liquid to water, box of paper clips, small, plastic cups (4 oz. or smaller)

Procedure 1. Water
1. Place a piece of paper towel under the small, 4 oz. plastic cup.
2. Pour water into the cup until the cup is full.
3. Add more water with an eyedropper until you see the water start to form a "dome" on the top of the cup.
4. Add paper clips one at a time by carefully and slowly sliding each clip into the cup.
5. Count the number of clips you can add before the water dome actually breaks and water flows over the side of the cup and wets the paper towel.

Procedure 2. Soapy Water (Water and Dishwashing Liquid)
Repeat the above procedures except add about one teaspoon of dishwashing liquid to the water. Gently mix the dishwashing liquid with the water so as not to create bubbles. If the water is bubbly, begin again with less dishwashing liquid.

Observations/Data: Class Chart
1. Your teacher will assign each group a name. Together, your class will design a chart that will record the number of paper clips each group added to their water and dishwashing liquid cups before the liquids overflowed.
2. Make a copy of the class chart in your notebook.
3. Record your findings in your notebook and on the class chart.
4. Graph your findings.

Claims and Evidence
1. How many paper clips were added to the water based on your group's results?
2. Compare your findings to the class results.
3. If someone in another class were to repeat this experiment, predict how many paper clips they might add before the water overflowed. Hint: Base your claim on evidence gathered from your data and class data.
4. How many paper clips were added to the soapy water based on your group's results?
5. Compare your findings to the class results.
6. If someone in another class were to repeat this experiment, predict how many paper clips they might add before the water overflowed. Hint: Base your claim on evidence gathered from your data and class data.
7. Which liquid, water or soapy water, was able to hold the most paper clips before spilling? State your claim and back it with evidence.
8. How does your prediction compare to your actual results?

New Understandings
1. Why do you think the water dome got larger and didn't overflow the cup as more paper clips were added?
2. Why do you think the experiment was done with both water and soapy water?
3. Why did groups get different results?
4. What did you learn that you did not know before?
5. What other questions do you have about "the behavior of water?"
6. What other liquids might be interesting to test and why?

Reflections
Comment about one of the following:
1. What changes would you make if you repeated this investigation?
2. How well did your group work together?
3. Would you recommend this investigation to future students?

Teacher Notes: How Many Paper Clips?
Lesson Two – Student Investigation

Rationale: This investigation is highly motivating and presents students with a discrepant event. Very few students predict the large number of paper clips that are added before the cup of water overflows. They are usually surprised by the results. As they compare the number of paper clips that can be added to a cup of water and a cup of soapy water, they become curious as to why water behaves as it does. This sets the stage for the study of surface tension.

Materials: Safety goggles, tray, paper towels, eyedroppers, toothpicks, 250 mL beaker, dish-washing liquid, plastic spoon, #2 paper clips, 4 oz. or smaller plastic cups

Preparation
1. Use small, plastic, 4 oz. cups. Paper cups tend to develop leaks. The larger the cup, the more paper clips you'll need; thus, 4 oz. cups are highly recommended. Make sure cups are plastic as paper cups frequently "come apart" or leak as more clips are added.
2. Use #2 sized paper clips in 4 oz. cups. Each group will need at least 100 clips.
3. Demonstrate how to set up the cup with a slight "water dome" before clips are added. Use an eyedropper to show how to add more water to create the dome.
4. Paper clip adding technique should be addressed before students collect data. Student groups should be cautioned to "slide" rather than "drop" the clips into the water so that the dome breaks due to the added number of clips rather than poor technique.
5. Rinse off soap and dry clips on paper towels between use and before storing for the next use.

Teacher's Role: The teacher's role is to organize the groups, demonstrate lab technique, manage the materials, monitor laboratory work, facilitate class data chart construction, and conduct a class discussion based on claims and evidence and new understandings.

At this time, the students should try to formulate their own definition for surface tension. Encourage students to use the word *surface tension* and explain it in their own terms. In the next lesson, the reading provides a formal explanation.

Background: Once students have collected their data, they will combine their team results in a class chart like the one shown below. This is an opportunity to discuss how multiple sets of data can be organized and compared. When students see the range of differences in the team findings, they usually suggest averaging the data. This is an excellent time to discuss measures of central tendency: mean, median, and mode. A sample chart is provided below with sample data.

Sample Data Chart: Water vs. Soapy Water		
Team	Water Clip Number	Soapy Water Clip Number
Team 1	95	65
Team 2	110	80
Team 3	120	75
Class Average	108	73

Claims and Evidence: Student answers will vary based on technique, size of cup, size of clips, and other variables. However, more clips will be added to the water than the soapy water.

New Understandings
1. **Why do you think the water dome got larger and didn't overflow the cup after many paper clips were added?** Hopefully, students will talk about the "skin" or "surface" of the water holding the water in the cup. They may even use the term "surface tension." It is not necessary to define surface tension until all students complete the reading that follows.
2. **Why do you think the experiment was done with both water and soapy water?** Students should recognize that the surface was not as "tight" with the dishwashing liquid. And, that the dishwashing liquid did something to the water's surface.
3. **Why did groups get different results?** This is a good time to review "sources of scientific error or uncertainty" for this experiment. Probable sources of error are variations in techniques like determining dome height, how clips are added, and how liquids are measured.

Teaching Tip
Extension: Students can try other liquid solutions (e.g., salt water, sugar water, and rubbing alcohol) and compare the results to water.

Lesson Three: Water's Amazing Invisible Elastic Skin
Student Reading

Wow! Who would have thought that you could add paper clip after paper clip to a full cup of water without the water overflowing? This is just one of the properties that makes water – WATER! Now it is time to explain the observations by applying what scientists already know about water.

Water remains in the cup after each paper clip is added because of the **strong cohesive attractive forces** between water molecules. **Cohesion is used in chemistry to describe the attractive force between molecules of the same substance.** The **cohesive forces** that hold water molecules surrounded by other water molecules together are **balanced forces**. In other words, **balanced, cohesive attractive forces** hold water molecules in water drops together.

To understand what is meant by a **balanced force** it might help to imagine that you are standing in the middle of a group of people. Imagine people pulling on you from all directions as you pull on them in all directions with the same force. The forces pulling on you are opposite but equal. The forces acting on you are **balanced.**

Surface water molecules are different. The water molecules at the surface of a liquid are not completely surrounded by other water molecules. They also contact air molecules above the surface. The air molecules pull on the surface water molecules with much less force than other water molecules. The forces acting on the surface water molecules are *unbalanced*, **cohesive attractive forces.**

To understand what is meant by *unbalanced* forces imagine you form a circle around a crowd of people by holding hands with the people next to you. The people in the center of the circle are pulling you inward. The people next to you are pulling you sideways. But, no one is pulling you outward. In this situation you are being pulled inward more strongly than outward. The forces acting on you are now *unbalanced* forces.

The *unbalanced* **forces acting on water molecules** at the water's surface cause water to act like it has an **elastic skin**. This *unbalanced force* at a **liquid's** surface is called **surface tension.**

Water's Amazing Invisible Elastic Skin

Student Post: Reading Activity

Summary Paragraph

Now that you have read about surface tension, it is time to summarize what you have read.

1. In your science notebook, write the paragraph below using the **highlighted** words and/or phrases from the reading to complete each sentence. **Note:** Not all words may be needed. You may need to make some words plural or singular. You may only use **highlighted** words/ phrases.

2. When you are finished, read your paragraph to another student or your teacher.

3. Revise your work, if needed.

Summary Paragraph

(1) _____ forces hold (2) _____ together. These (3) _____ forces are (4) _____ at the water's (5) _____. The surface (6) _____ are pulled more strongly toward other (7) _____ than molecules in the air above the surface. The surface (8) _____ act like an (9) _____. The (10) _____ of water is (11) _____ enough to form a dome or a rounded drop. Surface tension is a property of water and all other (12) _____.

Teacher Notes: Water's Amazing Invisible Elastic Skin
Lesson Three: Student Reading

Rationale: The reading provides students with a written explanation for the concept of surface tension. Before students can construct their own explanations for scientific phenomena, they need experience reading scientific explanations.

Literacy Strategies: QAR – Question-Answer Relationships; Everybody Read To…(ERT), and Summary Paragraph. See Literacy Resource Section.

Reading Strategies

Before Reading: Lesson Two – Investigation: How Many Paper Clips?
During Reading: QAR with Shared Reading: Everybody Read To…(ERT)
After Reading: Summary Paragraph

Background

Content: Cohesion, or the *inter*-molecular force holding molecules of the same substance together, is introduced in this reading. It will be contrasted with adhesion, the force holding molecules of different substances together in a later reading. Introducing contrasting terms like cohesion and adhesion separately is better practice than introducing them at the same time.

Teacher's Role: During Reading
QAR with Shared Reading: Everybody Read To…(ERT)

Paragraph One: Introduce the reading by asking students to read the title and the first paragraph. Tell students to raise their heads and look at you when they are finished reading the first paragraph. Wait until students are facing forward.

1. Prompt students to raise their hands if they can tell the **purpose** of the reading.
2. Wait until at least $1/3^{rd}$ of the class raises their hands and then call on students with raised hands.
3. Ask each student to read the part of the paragraph that tells the **purpose** of the reading. Accurate responses should state that the purpose is to explain observations about water. **From the Reading:** Now it is time to explain our observations by applying what scientists already know about water. Before reading further, initiate a discussion to review and determine what students already know about water's surface.
   a. How is an explanation different from an observation?
   b. Can you explain why water could hold more paper clips than soapy water?
4. Repeat the above procedure with different questions for each succeeding paragraph. You may create your own questions or use the QAR questions that follow.

Sample QAR Questions (Note: Some QAR questions can be answered directly from the reading, some require students to make inferences, and some require students to use their background and experiences.)

Paragraph Two QARs
1. **What is meant by the term cohesion?**
From the Reading: Cohesion is used in chemistry to describe the attractive force between molecules of the same substance.

2. **Are attractive forces stronger between the molecules in water or soapy water? Explain.**
Stronger in water because the dome got bigger and took more paper clips to break.

3. **What forces hold atoms of aluminum together to make aluminum foil?**
Cohesive forces exist between aluminum atoms.

Paragraphs Three QARs
1. **What is a balanced force?**
From the Reading: Students can read entire paragraph or paraphrase.

2. **Are water molecules moving in the cup?**
From Lesson Set One: Yes, liquid molecules are always in motion. They have kinetic energy. They vibrate, spin, and move around other molecules.

3. **How can a water molecule have a balanced force yet still have kinetic energy?**
Think of a water molecule like a person in a crowd. They are pushed around but remain in relatively the same place if the forces acting on them are balanced.

Paragraph Four and Five QARs
1. **How are surface water molecules different from the other water molecules inside the cup?**
From the Reading: The water molecules at the surface of a liquid are not completely surrounded by other water molecules. They are also in contact with air molecules above the surface.

2. **How are the forces acting on a water molecule in the center of a cup different from the forces acting on a water molecule on the surface of the cup?**
The forces acting on water molecules in the middle of the cup are balanced and the forces acting on the molecules on the surface are unbalanced.

3. **Both like and unlike atoms and molecules can be attracted to each other. Using this information, can you explain why a meniscus (U-shaped surface) forms when water is added to a graduated cylinder or tube?**
Water and glass molecules are strongly attracted to each other; thus, water molecules stick to the sides of the cylinder as the water molecules below the surface pull surface water molecules inward. The forces are unbalanced.

Paragraph Six QARs

1. Define "surface tension."

From the Reading: The *unbalanced* **forces acting on water molecules** at the surface.... is called surface tension. Note: In this case, a phrase from one sentence is added to a phrase from the last sentence. Students need to be taught that they may have to put different parts of text together to fully comprehend a nonfiction selection.

2. Other than adding clips to water, give another example of water's strong surface tension.

Depends on student's experience and background. Possible answers: seeing how many water drops can be placed on the head of a penny, floating a pin on water, an insect walking on water.

3. What liquids might have a weaker surface tension than water?

Depends on student's experience. Possible answers: soapy water, alcohol, liquids that are less dense.

4. What liquids might have a stronger surface tension than water?

Depends on student's experience. Possible answers: oils, dense liquids

Teacher's Role: After Reading

Your role is to guide students to use text features, like highlighted terms, to complete the summarizing paragraph. Students will need to use both words and phrases. They will also need to make some words plural.

Highlighted Reading Words and Phrases: Water's Amazing Elastic Skin (Title), water molecule, strong cohesive attractive forces, cohesion, balanced forces, unbalanced forces, elastic skin, and liquid's surface tension

Scoring Key for Summary Paragraph Note: There may be more than one-way to fill in the blanks for the paragraph to make sense.

(1) **Strong cohesive attractive** forces hold (2)**water molecules** together. These (3)**strong cohesive attractive** forces are (4)**stronger/unbalanced** at the water's (5)**surface.** The surface (6)**water molecules** are pulled more strongly toward other (7)**water molecules** than molecules in the air above the surface. The surface (8) **water molecules** act like an (9)**elastic skin**. The (10)**surface tension** of water is (11)**strong** enough to form a dome or a rounded drop. Surface tension is a property of water and all other (12)**liquids**.

Lesson Four: Explore Some More – Sinkers and Floaters
Student Investigation

Introduction

Surface tension causes water to behave as if it has an invisible skin. Water's invisible elastic skin allows objects that are denser than water to float when they should sink. For example, a leaf that is denser than water may float on the surface of a pond until the wind or other force causes it to sink. Can you think of other objects that might both float and sink in water?

Focus Questions

1. What common objects can both float and sink in water?
2. What are the characteristics of objects that both float and sink in water?

Procedures & Observations

Design your own procedures to gather evidence to support your answers to the focus questions.

Hints: As you added paper clips to your cup of water, you may have unexpectedly floated one on the water's surface, especially if you gently placed the clip in the water with very little force. Of course, if you drop a clip into the water, the clip usually sinks. Thus, a paper clip may be an object that you can get to both sink and float in water. Listed in the materials available are some objects that **might** both sink and float. It is okay to try objects other than those listed in the materials available, as long as you get permission from your teacher.

Design a technique to place objects on the water's surface as gently as possible. Remember, you can't claim an object will both sink and float unless you personally observe it floating and sinking.

Determine the data (measurements) you need to collect for each object.

Materials Available: Safety goggles, balance, graduated cylinder, metric ruler, Petri dish or shallow bowl, water, paper towels, calculator, test objects (e.g., paper clip, needle, plastic objects, rubber objects, glass objects, coin, sand, wood objects, plant parts…)

Claims and Evidence

New Understandings

Reflections

Teacher Notes: Explore Some More – Sinkers and Floaters
Lesson Four – Student Investigation

Rationale: This is an opportunity for students to design their own investigation and investigate the difference between surface tension and density.

Materials: Safety goggles, calculator, balance, graduated cylinder, metric ruler, Petri dish or shallow bowl, water, paper towels, objects to test (e.g., paper clip, needle, plastic objects, rubber objects, glass objects, coin, sand, wood objects, plant parts…)

Materials Preparation
Prior to class, collect objects for students to test. Students may not understand the reason for the calculator, balance, ruler, and cylinders. These are included in the list to get them thinking about what type of data to collect and measurements to make.

Teacher's Role
The role of the teacher is to facilitate student investigation by managing materials and monitoring student work. The teacher will also use inquiry-based questioning techniques to focus student learning and thinking. The teacher should answer only the content questions that cannot be answered from other sources available to the students. This is the student's time to learn perseverance and persistence and apply research skills. Of course, any questions relating to safety should always be answered directly.

Because this lesson uses simple, easy-to-obtain, and safe materials, students can work in groups or work alone.

Samples of Teacher-Posed, Inquiry-based Questions
1. Why do you think some objects can both sink and float in water?
2. Besides observing whether or not an object sinks and floats, what other data should you collect?
3. What scientific concepts explain the behavior of the objects?
4. Is there another way you could approach this investigation?
5. Do you have enough evidence to make your claims?
6. What material is that object made from? Where could you find that information?
7. What properties of liquids relate to the sinking and floating?

When students finish with their investigation, the teacher's role is to debrief the learning. Students engaged in self-directed inquiry are usually very eager to share their findings and discussions are usually lively and exciting. Although this is the goal of every class discussion, it is hard to achieve. Besides telling what objects both float and sink, students should be asked about their thinking and feelings as they struggled to design their procedures and write their claims and evidence. Sometimes, students will express feelings of frustration and uncertainty as well as joy and satisfaction. It is important to tell students that scientists are no different. If students are not involved in the inquiry process, they will not understand how the joy of discovery learning is a key motivator for both students and scientists.

Background

Objects that both float and sink in water will have a density greater than one. The concept of density is not mentioned on the student pages. Because density was the focus of Lesson Set Two, this lesson requires that the students apply previous information in a new situation. The hope is that they will come up with the idea of comparing the density of the objects to the density of water on their own.

The density of water is 1 g/cm^3. Although students should know this, some may need to revisit the methods for calculating density. Students may not have to directly calculate the density of objects like a sewing needle. They can use references to find out that needles are made of a steel wire (iron alloy) and usually plated with nickel. Using the density of iron as a standard, they could conclude that the needles, although small and thin, have a density of about 7 g/cm^3. By comparing this to water's density (1 g/cm^3), they will note that a needle will sink if the strength of the surface tension is broken.

Claims and Evidence

Objects that can be supported by water's surface tension and sink in water are made of substances with a density greater than one. They also have to have a shape or design that allows their weight to be distributed over the surface of the water like a thin needle, flat leaf, or extended paper clip. Of course, many denser than water objects will never float (e.g., lead weight) because their surface pressure overwhelms the surface tension.

Teaching Tip

This is an excellent investigation to use to assess science process skills. Carefully observe student work habits and science process skills during the lesson. Make notes to assist in your evaluation of their learning. Reinforce good work habits and scientific thinking shown by individual students and groups.

Lesson Five: Concept Map Master – Surface Tension
Student Literacy

Directions: Copy the concept map below in your science notebook. Use the information from the reading *Surface Tension: Water's Amazing Invisible Elastic Skin* and your investigations to complete the map.

Bigger Idea Here

Surface Tension

Give Examples of Surface Tension

List Properties of Surface Tension

Teacher's Notes: Concept Map: Surface Tension
Lesson Five: Student Literacy

Rationale: Teaching science is about teaching concepts. Concept maps help students integrate and synthesize information for deeper understanding of key ideas.

Materials: Student copy of concept map master.

Teacher's Role: If students have not previously used a concept map, model for the students how to complete a concept map using a simple concept. For example, replace the word "surface tension" with a familiar animal like a dog. Have students tell you how to fill in the squares and lines as you guide them through the process.

Background: Below are sample student responses for "Surface Tension."

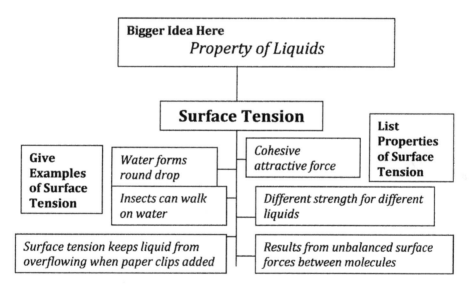

Teaching Tips

- Concept maps should be done individually and assessed for accuracy by the teacher.

Lesson Six: Observing Water Drops
Student Investigation

Introduction: Attractive forces exist between *like* molecules. Attractive forces also exist between *unlike* molecules. In this investigation, you'll study the attractive forces between water molecules and other molecules. To do this, you'll place water drops on different surfaces and observe the shape and behavior of the drops. You will test surfaces like, plastic, metal, glass, wood, paper, and fabric.

Focus Question: How does the shape of a water drop relate to its attraction to different surfaces?

Prediction:

Materials Per Group: Safety goggles, toothpicks, metric ruler, cup of water, eyedropper, paper towel for clean up, Test surfaces: plastic wrap, wax paper, aluminum foil (or other metal surface), glass plate, wood, tile, paper towel, magazine cover, plastic plate, fabric, leaf.

Procedures
1. Place a piece of wax paper on the work surface.
2. Using an eyedropper place a few drops of water on the test surface.
3. Observe and draw the drops from the both a side view and from a top view.
4. Using the centimeter (cm) ruler, record the height and diameter of a typical drop in millimeters (mm).
5. Use a toothpick to try to drag the drops along on the test surface. Record your findings.
6. Repeat steps 1-6 using other available surfaces.

Observations
1. For each surface, draw a side and top view of the drop.
2. Use diagrams, words, and measurements to record your observations.
3. Make sure all parts of your drawings are labeled.

Claims and Evidence
1. To which surfaces are water drops most attracted? Support your claim with evidence.
2. Can you make any other claims about the behavior of water drops on different surfaces?

New Understandings
1. Why do you think water drops act differently on different surfaces?
2. Why is it important for scientists to record careful, systematic observations?
3. What other surface(s) would you like to test that you did not get a chance to test? Why?

Reflections
1. If you could describe this lab with one word or phrase, what would it be and why?
2. Other ideas, comments, or suggestions?

Teacher Notes: Observing Water Drops
Lesson Six: Student Investigation

Rationale: Using familiar materials, students study adhesive attractive forces between different substances. By placing water drops on wax paper, plastic wrap, and other surfaces, students observe the effect of adhesive attractive forces.

Note: This lesson is similar to Matter's Phases, Lesson Six: Comparing Water's Properties to Other Liquids. In that lesson, students placed water, mineral oil, rubbing alcohol, and vinegar on both wax paper and plastic wrap. The purpose of the Matter's Phases lesson was to distinguish between liquids by observing the shape of the liquid's drops on wax paper and plastic wrap. The purpose of this lesson is to determine the strength of adhesive forces by observing water's shape on a variety of different surfaces.

Materials: Safety glasses, piece of wax paper (10 cm x 10 cm), piece of plastic wrap (10 cm x 10 cm), toothpick, metric ruler, cup of water, eyedropper, paper towels, and other test surfaces like: aluminum foil (or other metal surface), glass plate, wood, tile, paper towel, magazine cover, plastic plate, fabric, leaf.

Preparation and Procedures: Prior to the lesson, collect various materials for students to use as surfaces for the water drops. Allow students to select their "other" surfaces from those available. Each group should test 3-5 surfaces.

For this investigation, record measurements in millimeters (mm). Always use metric units when available and appropriate. Students learn metrics best by using it frequently and correctly. Make sure they use correct unit measurements and their abbreviations.

Teacher's Role: The teacher's role is to manage materials and monitor student behavior. When the lesson is complete, the teacher facilitates a class discussion based on stated claims and evidence as well as new understandings.

Background
The attraction of unlike molecules to each other is called adhesion. Adhesive forces will be discussed further in the next lesson.

Claims and Evidence
1. **To which surfaces are water drops most attracted? Support your claim with observational evidence.** Water will bead on surfaces that have a weak attraction to water like wax paper. Water will spread out on surfaces with a strong attraction to water like plastics.
2. **Can you make any other claims about the behavior of water drops on different surfaces?** Probable observations are: 1) some surfaces absorb the drops, 2) when drops are placed close together and touch, they coalesce to form one big drop, 3) some surfaces allow drops to be pulled with a toothpick and other surfaces do not, 4) some drops cling to surfaces

even when the surface is tilted vertically, and 5) small drops cling to vertical surfaces and do not run down.

New Understandings

1. **Why do water drops act differently on different surfaces?** Attractive forces exist between unlike molecules. These forces can be weaker or stronger depending on the type of molecules. Strong attractive forces exist between water and plastic wrap molecules; thus, the molecules spread or flatten out. Weak attractive forces exist between wax paper molecules and water; thus, the water molecules are not attracted to the wax molecules and bead up.

2. **Why is it important for scientists to record careful, systematic observations?** This question directs students to think about scientific observations as different from other types of observations.

Reflections

1. **If you could describe this lab with one word or phrase, what would it be and why?** This type of question is designed to draw from the student's background information to connect new learning with previous learning. Possible word/phrases are: clinging drops, sliding drops, sticky drops, pulling and pushing water, combining drops, etc. As students use words and phrases, follow up with "Why did you choose those words?" The words or phrases are prompts to evoke more detailed responses.

Lesson Seven: May the Forces be With You
Student Reading

You just observed water beading on a waxed surface and spreading on a plastic surface. A great question is, "Why doesn't water bead on plastic wrap or spread on waxed paper?" Here's what scientists think.

Cohesive, attractive forces hold water molecules together. **Cohesion** is the word scientists use to describe the **attractive forces that hold the same kind of molecules or atoms together**. But, molecules of different substances can also be attracted to each other. **Adhesion** is the word used to describe **the attractive forces that hold different kinds of molecules and atoms together**, like water molecules and plastic wrap molecules.

When water is dropped on wax paper, it beads because water molecules and wax molecules do not have a strong, **adhesive** attraction to each other. In this case, water molecules have a stronger, **cohesive** attraction to each other. Thus, the water does not spread out.

Cohesive and adhesive forces can be weak or strong. The cohesive forces between water-to-water and wax-to-wax molecules are strong. The adhesive forces between water-to-wax and wax-to-water molecules are weak. The adhesive forces between water-to-plastic wrap molecules are stronger. This is evidenced by water spreading out on plastic wrap and beading on wax paper. Water **adheres** better to plastic wrap than to wax paper.

Cohesive and adhesive attractive forces act on all substances. The strength of these forces will determine how substances behave when they come together. For example, if liquid water comes in contact with a paper towel, strong adhesive forces cause the water to **adhere** to the paper towel and become absorbed by the towel. Wringing a towel forces some of the water out by breaking the adhesive forces between the paper and the water.

Lesson Seven: Student Main Idea Map Master
Reading Strategy

Title of the Article:

Main Idea:

•**Important Idea:**
   ~ **Supporting details**

•**Important Idea:**
   ~ **Supporting details**

•**Important Idea:**
   ~ **Supporting details**

Summary Statement:

Teacher Notes: May the Forces be With You
Lesson Seven: Student Reading

Rationale: The reading provides students with a written explanation for the concepts of cohesion and adhesion. Before students can construct their own explanations for scientific phenomena, they need to read explanatory text.

Literacy Strategy: Main Idea Map. See Literacy Resource Section.

Reading Strategies
   Before Reading: Lesson Six – Investigation: Observing Water Drops
   During Reading: Main Idea Map
   After Reading: Main Idea Map

Materials: Student Main Idea Map Master, adhesive tape or duct tape

Teacher Demonstration. A quick demonstration can help students distinguish between adhesion and cohesion. Show a piece of adhesive tape to the class. Stick it to a surface and indicate that it is adhesive, because it adheres to a different type of surface. Now stick the tape to itself. The tape coheres to itself. It is sticking to the same surface. Duct tape also works well as a visual.

Teacher's Role: Main Idea Map with Bullet Meatballs and Spaghetti Dashes
Main Idea Maps are designed for students to "outline" the key concepts in content readings.
1. First, have students skim and scan the article to determine the article's main idea. Tell students that the **main idea** is the **most important idea** expressed in the article. When they finish, have them raise their hand. When at least $1/3^{rd}$ of the class has their hands raised, ask one student for a response. Check to see if others in the class agree. Resolve any differences and write the main idea on a board or overhead for everyone to see.
2. Tell the students that the next step is to read the article paragraph by paragraph to find **important ideas that support the main idea**. These ideas area are designated with a "•" bullet or meatball on their map. Using the term "meatball" helps students understand that important ideas are the meat of the article. Have students carefully read the first two paragraphs to determine whether or not there is a "meatball" or supporting important idea. When $1/3^{rd}$ of the class has their hands raised signaling they have read the paragraphs and are ready to report, ask one student for a response. Check to see if others in the class agree. Resolve any differences and write the "•" on the board or overhead.
3. Next, tell the students to write a "~" or "spaghetti" symbol under the bullet. Tell them that the spaghetti symbol indicates the details that support the important idea. The number of "~" symbols varies based on the number of supporting details provided in the article.
4. On a board or overhead, add at least one supporting detail from the reading.

5. Have the students continue reading to determine additional important ideas and supporting details. As students read and complete their maps, walk around the room and check for student understanding.

6. Lastly, as students finish, place them in groups of two or three and have them compare their outlines. Then, tell them to compose a summarizing statement. Have students share their summarizing statements with the class.

Sample Main Idea Map

Main Idea: *Attractive forces hold matter together*

•**Important Idea:** *Cohesion*
   –**Supporting details**
       ~ *attractive force*
       ~ *holds the same kind of molecules together*
       ~ *holds water molecules together in drops*
       ~ *creates surface tension*

•**Important Idea:** *Adhesion*
   –Supporting details
       ~ *attractive force*
       ~ *holds molecules of different kinds of molecules together*
       ~ *holds water molecules to plastic wrap molecules*

•**Important Idea:** Cohesion and Adhesion
   –**Supporting details**
       ~ *forces are different strengths and can be weak or strong*
       ~ *forces exist between all forms of matter*
       ~ *forces hold molecules together*

Summarizing Statements: *The attractive forces that hold atoms and molecules together are cohesion and adhesion. Cohesion holds like atoms and molecules together. Adhesion holds unlike atoms and molecules together.*

Lesson Eight: A Detergent's Peppery Power
Student Investigation

Focus Question: What is the effect of liquid dish detergent on the surface tension of water?

Prediction:

Materials Per Group
    Safety goggles
    Pepper (shaker)
    Liquid dish detergent (~5 mL)
    Water
    Aluminum pie plate
    Toothpicks (2-3)
    Wax paper (small square)
    Eyedropper
    Plastic punch cups (to hold water and detergent)
    Paper towels

Procedure 1. Water, Wax Paper, and Detergent
1. Place a large drop of water on wax paper.
2. Dip a toothpick in detergent and add it to the top of the water drop.

Observation: What happened when the detergent was added to the drop of water?

Procedure 2. Water, Pie Plate, and Pepper
1. Put enough water in the aluminum pie plate to cover the bottom.
2. Sprinkle pepper over the surface.
3. Dip a toothpick in the liquid dish detergent solution and then gently touch the toothpick tip to the middle of the pie pan.

Observations: What happened to the pepper when the liquid dish detergent touched the water?

Claims and Evidence: What evidence do you have that liquid dish detergent interferes with water's surface tension?

New Understandings
1. Pepper is denser than water. It should sink when added to water. How do you explain the observation that it floated on the surface of water?
2. How was the pepper test and water drop test the same? Different?

Teacher Notes: A Detergent's Peppery Power
Lesson Eight: Student Investigation

Rationale: This lesson engages students to learn more about surface tension and sets the stage for the next lesson.

Materials: Safety goggles, pepper, small beakers, liquid dish detergent, aluminum pie pan, toothpicks, wax paper, eyedropper, paper towel

Purchasing and Preparation

Aluminum pie pans are inexpensive and work well. If you use another container, it should be edged and shallow. The pie plate's large surface area allows for a dramatic effect when the soap drop is added.

For the pepper, one shaker can be used for an entire class. Since pepper can trigger sneezing, it is best to have the teacher shake the pepper onto the surface for the students.

Teacher's Role: The role of the teacher is to manage the materials and monitor student behavior. In the follow-up discussion, the teacher should not offer an explanation for the student observations. This investigation is designed solely to engage students and prepare them for the next two lessons.

Background

Liquid dish detergent is a surfactant. When added to water, it reduces the cohesive forces between water molecules and "breaks" the surface tension. Pepper is both a floater and sinker. That means that it is denser than water; however, the surface tension of water allows it to stay on the surface. The pepper sinks when the surface tension breaks.

Teaching Tips

- Students may explore further by finding other substances that float on water and move by adding a surfactant. They may want to try some of their "floaters and sinkers" from Lesson Four.
- Small, two-dimensional paper shapes can be cut and raced across the surface of a long, aluminum cake-type pan. Paper folded in the shape of a boat is fun!

Lesson Nine: Explore Some More – Breaking the Surface
Student Investigation

Introduction: A **surfactant** is a substance that has the ability to lower the surface tension of liquids. The most common example of a **surfactant** is liquid dish detergent. Liquid dish detergent molecules are able to break the **strong, cohesive, attractive forces** between water molecules. When a drop of detergent is placed on the surface of water sprinkled with pepper, pepper rapidly spreads out from the center. The detergent molecules interfered with the strong, cohesive attraction between the water molecules.

Focus Question: What effect will a surfactant have on milk?

Prediction:

Materials per Group
   Safety goggles
   Whole milk: Whole milk is a mixture of water, fat, sugar, protein, vitamins, and minerals.
   Reduced Fat Milk: 2% milk, 1% milk, skim milk
   Petri dish or similar container
   Red, green, blue, and yellow food coloring in dropper bottles: Food coloring consists of a
      dye dissolved in water.
   Liquid dish detergent in small beaker
   Toothpick/eyedropper
   Paper towels

Procedure
1. Fill a Petri dish or small, shallow bowl about ½ full of whole milk.
2. Place a few drops of several colors of food coloring on the surface of the milk. Note: The food coloring will help you observe any movement in the dish.
3. Using a toothpick or eyedropper, add a few drops of liquid dish detergent to center of dish.
4. Observe and repeat steps 1 – 3 with other types of milk.

Observations: Record in your science notebook.

Claims & Evidence
1. What evidence do you have that liquid dish detergent acts as a surfactant with milk?
2. What evidence do you have that liquid dish detergent is acting on more than just the water molecules in milk? Hint: How does the movement of the food coloring differ from the movement of the pepper?

New Understandings
1. What did you learn that you didn't know before?
2. What questions do you have about molecular attractive forces?

Teacher Notes: Explore Some More – Breaking the Surface
Lesson Nine: Student Investigation

Rationale: Students further explore the concepts of attractive forces, surfactants, and surface tension by performing a fun and novel experiment. This investigation and Lesson Eight, are the before reading activities and strategy for the explanatory reading in Lesson Ten.

Materials: Safety goggles, whole milk, 2% milk, 1% milk, skim milk, Petri Dish or similar container, food coloring in dropper bottles, liquid dish detergent in small beaker, toothpick/eyedropper, paper towels

Purchasing and Preparation
Whole milk, 2% milk, 1% milk, and skim milk. You do not have to use all four types of milk, if cost is an issue. You will need at least two different kinds. Since the swirling effect comes from the surfactant interacting with the fat and water in the milk, it is recommended that whole or 2% be compared to 1% or skim. Buttermilk is dense and thick and little swirling occurs. Students generally think that the more fat, the better. Not in the case of buttermilk. The next reading discusses homogenized milk.

Food Coloring. This lab is more dramatic when students use a variety of food colors. Be prepared for students to want to repeat this activity many times. It is a middle school favorite.

Teacher's Role: The teacher's role is to manage materials and monitor student work. Once students see the swirling, they almost always want to explore further. As you walk about the room, call student attention to some of the more spectacular swirling. Before reading the article in Lesson Ten, conduct a discussion to uncover student ideas; however, refrain from completing the explanation until students have done the Lesson Ten reading.

Background: See the reading in Lesson Ten for an explanation. Content Note: Whole milk is a mixture of water, fat, sugar, protein, vitamins and minerals. Food color consists of a dye dissolved in water. In this lesson, the surfactant both reduces the surface tension and adheres to the fat molecules in milk.

Claims & Evidence
1. **What evidence do you have that liquid dish detergent acts as a surfactant with milk?** The food color behaves similar to the pepper in the previous lesson.
2. **What evidence do you have that liquid dish detergent is acting on more than just the water molecules in milk?** The food coloring moves from the center to the sides of the dish, appears to sink and then resurfaces as it swirls. And, it behaves differently depending on the fat content in the milk. It seems to be interacting with the fat.

Teaching Tip: If a camera and computer are available, it is fun to take and download swirl pictures for students to see and even use as screen savers!

Lesson Ten: From Milky Swirls to Washing Clothes
Student Reading: Anticipation Guide

I. Read each of the following statements and determine if you agree or disagree.

A or D   1. Whole milk is about 10% fat.

A or D   2. A surfactant lowers the surface tension of water.

A or D   3. Dish detergent molecules cohere to both water and fat.

A or D   4. If milk is not homogenized, a creamy layer of fat will form on the bottom of the bottle or carton.

A or D   5. Food coloring will diffuse throughout whole milk.

A or D   6. Detergents adhere to grease in wash water.

A or D   7. Skim milk will swirl more than whole milk when food coloring and dish detergent are added because skim milk has less fat.

A or D   8. If whole milk is warmed, the food coloring should swirl faster.

II. With a partner, discuss your responses. You may change your answers before you do the reading if you wish.

III. Read the article, "From Milky Swirls to Washing Clothes". As you read, make any corrections to the statements. Be prepared to back up your answers with evidence from your reading.

IV. Discuss the statements again with either your partner and/or your teacher.

V. Summarize the important ideas and key words that relate to chemistry in your science notebook.

Lesson Ten: From Milky Swirls to Washing Clothes
Student Reading

Whole milk is about 4% fat. Fats and oils do not dissolve in water. Milk straight from the cow separates into two layers: cream on the top and less fatty milk on the bottom. When whole milk is readied for sale, it goes through a process called **homogenization. Homogenization uniformly suspends the fat molecules in the milk so that they don't separate.**

Because milk contains a high percentage of water, adding detergent to whole milk lowers the **surface tension** of milk just like it does for water. The food coloring moves away from the dish's center much like the pepper moved in water. At the same time, the liquid dish detergent surrounds and dissolves the milk's fat molecules.

Usually, substances are either **soluble** (able to dissolve) or **insoluble** (not able to dissolve) in water. However, detergent molecules are different. Detergent molecules **adhere** to both water molecules and fat molecules. Although we can't see a detergent molecule, it is known to be a long molecule that has a water-attracting end and a water-repelling tail. The water-repelling tail adheres to fat molecules. Thus, detergents can dissolve in water and in fat at the same time!

As **the detergent** diffuses through the milk, the food coloring molecules are bombarded from all sides creating fascinating color swirls. Some factors that affect swirling are the temperature of the milk, the amount of food coloring, the amount and kind of surfactant, as well as the amount of fat in the milk.

Clothes washing detergents remove grease and grime from clothing by adhering to fat molecules on fabric and water molecules. During the rinse cycle, the grease-detergent-water molecules are removed from the wash water as it drains.

Teacher Notes: From Milky Swirls to Washing Clothes
Lesson Ten: Student Reading

Rationale: The reading explains the concepts underlying the milky swirls investigation. It reinforces the concepts of adhesive forces and surface tension while reinforcing the concept of diffusion.

Strategy: Anticipation Guide: See Literacy Resource Guide.
   Before Reading: Lessons Eight and Nine, Anticipation Guide Parts I & II.
   During Reading: Anticipation Guide Part III.
   After Reading: Anticipation Guide Parts IV and V.

Materials: Student Copy of Anticipation Guide and Reading

Teacher's Key: Anticipation Guide – From Milky Swirls to Washing Clothes
I. Read each of the following statements and determines if you (A) agree or (D) disagree.

A or **D**   1. Whole milk is about **10% fat. Disagree. It is 4% fat.**

A or D   2. A surfactant lowers the surface tension of water. **Agree**

A or **D**   3. Dish detergent molecules **cohere** to both water and fat. **Disagree. They adhere not cohere.**

A or D   4. If milk is not homogenized, a creamy layer of fat will form on the bottom of the bottle or carton. **Disagree. Cream forms on top.**

A or D   5. Food coloring diffuses throughout whole milk. **Agree. Note: The food coloring diffuses faster with the help of the surfactant. It will diffuse without the surfactant, but it will take longer.**

A or D   6. Detergents adhere to grease in wash water. **Agree.**

A or **D**   7. Skim milk will swirl **more** than whole milk when food coloring and dish detergent are added because skim milk has less fat. **Disagree. Less fat, less swirls.**

A or D   8. If whole milk is warmed, the food coloring should swirl faster. **Agree.**

Lesson Eleven: Surface Tension Measurements
Student Chart Interpretation

Introduction: The surface tension of liquids can be measured in a laboratory. Listed below are the actual surface tensions of some common liquids. The measurements are in mN/m or milli-Newton's per meter. A Newton is a measure of force. Gravity's force on a person weighing about 155 pounds is approximately 687 Newtons. A milli-Newton is one-thousandth of a Newton. Thus, a 155 pound person would weigh 687,000 milli-Newtons (1000 x 687)!

Surface Tension of Liquids at Specific Temperatures		
Liquid	**Temperature °C**	**Surface Tension mN/m**
Water	1	75.6
Water	20	72.8
Water	25	72.0
Water	50	67.9
Water	100	58.9
Acetone (Nail Polish Remover)	20	23.7
Ethanol	20	22.3
Glycerol	20	63.0
Isopropanol (Rubbing Alcohol)	20	21.7
Salt Water (35% solution)	20	82.6
Sugar water (55% solution)	20	76.5

Interpreting the Chart
1. What happens to the surface tension of water as the temperature rises?
2. What happens to the surface tension of water when salt is added?
3. What substance has the lowest surface tension at 20°C?... The highest surface tension?
4. Which substances at 20°C have a surface tension strong enough to "float" a pin? Hint: Under what conditions did you float a metal object?
5. Which substances have strong, cohesive attractive forces between their molecules? Which have weak, cohesive attractive forces? Hint: Water has strong attractive forces.
6. The chart above has the surface tension of sugar water. What would be the surface tension of solid sugar?
7. What else can you learn from the chart? What would you like to learn?

Teacher Notes: Surface Tension Measurements
Lesson Eleven – Student Chart Interpretation

Rationale: Chart reading is a learned skill that requires practice reading and interpreting numerical data. In this lesson, quantitative information is used to compare the surface tension of water to other liquids and to learn how temperature affects surface tension.

Teacher's Role: The role of the teacher is to monitor student work and check for student understanding. Students should work both alone and collaboratively to answer the questions.

Background: Surface tension is a property of liquids. It results from the unbalanced, cohesive attractive forces between molecules at the surface of a liquid. It is reported in mN/m or milli-Newtons per meter. Surface tension is a characteristic property of liquids. Water has a strong surface tension. Other liquids like acetone and alcohol have weaker surface tensions. Liquids with weak surface tensions evaporate quickly and have lower boiling points.

Temperature increases the kinetic energy of molecules. As temperature increases, surface tension decreases. Molecules are moving faster and have more kinetic energy. When temperature rises the cohesive forces can still hold the molecules in contact with each other, but not in a fixed shape. Adding salt and sugar to water increases water's surface tension. Salt is an ionic molecule and sugar is a polar molecule.

Interpreting the Chart
1. **What happens to the surface tension of water as the temperature rises?** The surface tension of water decreases as its temperature rises. As the temperature rises from 1°C to 100°C the surface tension decreases from 75.6 mN/m to 58.9 mN/m, a decrease of 16.7 mN/m.
2. **What happens to the surface tension of water when salt is added?** Since water's surface tension is 72.8 mN/m at 20°C and salt water's surface tension is 82.6 mN/m, the salt water's surface tension will be higher.
3. **What substance has the lowest surface tension at 20°C?** Isopropanol has the lowest surface tension at 21.7 mN/m. **Highest surface tension?** The salt water solution has the highest surface tension at 82.6 mN/m.
4. **Which substances at 20°C have a surface tension strong enough to "float" a pin?** Since the temperature of our room was about 25°C for water when a pin floated, substances that have a surface tension about the same or greater than water (72.0 mN/m or higher) will most likely float a pin. These would be water at 25°C or lower, the salt water solution, and the sugar solution. The others might float a pin but we would need more information.
5. **Which substances have strong, cohesive attractive forces between their molecules?** Water, glycerol, sugar water, salt water. **Which have weak, cohesive attractive forces?** Ethanol, acetone, and isopropanol. High surface tension equals strong cohesive forces between molecules.

6. **The chart above lists the surface tension of sugar water. What would be the surface tension of solid sugar?** Surface tension is a property of liquids, not solids.

7. **What else can you learn from the chart? What would you like to learn?** Answers will vary.

Teaching Tips

- Discuss the liquids before the students begin working. All are found in the supermarket or easily obtainable. Ethanol is also known as drinking alcohol. Use an alcohol thermometer as an example of ethanol. Glycerol is known as glycerin. It is also found in the pharmacy and has multiple uses for easing insect bites and soothing burns. A small amount of glycerol can also be added to liquid dish detergent to make a great bubble solution.

- **Graphing Extension**: Since many national, state, and local assessments require students to graph scientific data, this lesson could be extended to have students graph the data before answering the questions. The surface tension can be plotted against the temperature of water for a line graph and/or students can make a bar graph comparing the surface tension of liquids at 20°C.

Graphs can be checked for the following: 1) Title expressing both variables, 2) Dependent variable (surface tension) on y-axis properly labeled, 3) Independent variable (temperature) on x-axis properly labeled, 4) Correct type of graph: line or bar graph – line graph for continuous data and bar graph for discontinuous data, 5) Appropriateness of scale and utilization of space, and 6) Accuracy of plots.

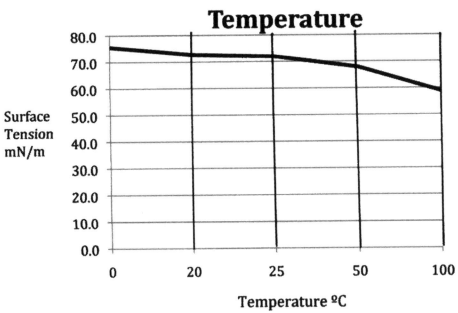

Surface Tension of Liquids

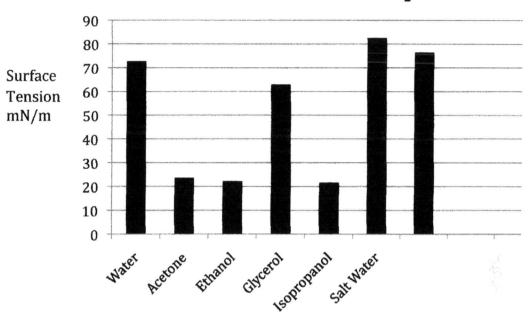

Lesson Twelve: Cohesion, Adhesion, or Both?
Student Review

Directions: Read each of the sentences below. In your science notebook, write the number of the sentence and then tell whether it describes an example of adhesion or cohesion and why. Add any additional comments and/or questions. When you have finished, pair-share your work with a partner.

Statements

1. Tires provide cars with traction on roads.

2. When syrup is poured from the bottle, it flows slowly.

3. A sticky note is posted on a wall.

4. Sugar forms sugar crystals.

5. Sugar dissolves in coffee.

6. Oil spilled from a tanker floats on the surface of the ocean.

7. When milk is spilled, a sponge cleans up the mess.

8. A child adds together pieces of play dough to make an animal sculpture.

9. Rice sticks together after it is cooked.

10. When it rains, water beads on the newly waxed hood of a car.

Teacher Notes: Cohesion, Adhesion, or Both?
Lesson Twelve: Student Review

Rationale: Students review concepts to prepare for the Post-assessment. In all cases, students must apply what they have learned in new situations deepening their conceptual understandings.

Teacher's Role: The questions can be used in multiple ways: class discussion guide, collaborative group work, or homework with class discussion.

Background
1. **Tires provide cars with traction on roads.**
Adhesion – Tire or rubber molecules adhere to the road surface

2. **When syrup is poured from the bottle, it flows slowly.**
Cohesion – Syrup molecules are strongly attracted to each other.

3. **A sticky note is posted on a wall.**
Adhesion – The glue molecules on the sticky note are attracted to the wall molecules.

4. **Sugar forms sugar crystals.**
Cohesion – sugar molecules are attracted to each other.

5. **Sugar dissolves in water.**
Adhesion – sugar molecules are attracted to water molecules.

6. **Oil spilled from a tanker floats on the surface of the ocean.**
Cohesion – Oil molecules strongly cohere to each other. Water molecules strongly cohere to each other. Oil and water molecules have a very weak attraction to each other and do not mix. Thus, they are insoluble.

7. **When milk is spilled, a sponge cleans up the mess.**
Adhesion – Sponge molecules adhere to milk molecules.

8. **A child puts together pieces of play dough to make an animal sculpture.**
Cohesion – play dough molecules are attracted to each other.

9. **Rice sticks together after it is cooked.**
Cohesion – rice molecules are attracted to each other.

10. **When it rains, water beads on the newly waxed hood of a car.**
Cohesion and Adhesion – *cohesive forces* between water molecules are stronger than the weak *adhesive forces* between the water molecules and wax molecules of the car.

Lesson Thirteen: Attractive Forces Post-Assessment
Student Post-Assessment

Part 1. Agree or Disagree

Read each statement. Determine if you Agree (A) or Disagree (D) with the statement. If you disagree, tell why you disagree below the statement.

1. _____ As more and more paper clips are added to an already full cup of water, the water dome on the top of the cup will get larger and larger.

2. _____ Cohesion describes the attraction between unlike molecules. For example, cohesive forces cause water to spread out on plastic wrap.

3. _____ Adding paper clips to a full cup of water created an upward force that eventually broke the water's surface tension, and water overflowed the cup.

4. _____ Liquid dish detergent increased the surface tension of water.

5. _____ Temperature can affect surface tension.

6. _____ Surface tension is a characteristic of solids, liquids, and gases.

7. _____ Water molecules strongly adhere to wax molecules.

8. _____ Detergent molecules adhere to both water and fat molecules.

9. _____ Forces on molecules at the surface of a liquid are balanced as compared to forces on water molecules in the middle of a liquid.

10. _____ Attractive forces between molecules can explain the behavior of matter.

Part 2. Multiple Choice

Directions: Write the letter of the "best" answer. If you wish to explain why you chose that answer, write your response by the question.

_____1. Surface tension is a characteristic property of
  a. only solids.
  b. only liquids.
  c. only gases.
  d. solids, liquids, and gases.

_____2. Cohesion is best described as the
  a. attraction between pepper and water.
  b. force that holds molecules of the same substance together.
  c. sticking together of two different substances.
  d. way that liquid molecules move about in liquids.

_____3. A surfactant, like dishwashing liquid,
  a. reduces surface tension.
  b. increases surface tension.
  c. preserves substances.
  d. cools substances.

_____4. Which is the best explanation for why oil and water do not mix?
  a. Oil particles are slippery.
  b. Water particles are only attracted to other water particles.
  c. Oil is denser than water.
  d. Oil particles are weakly attracted to water particles.

_____5. When a liquid is cooled, the surface tension
  a. increases.
  b. decreases.
  c. remains the same.
  d. cannot be measured.

_____6. Attractive forces between water molecules and plastic wrap are best described as
  a. cohesive forces.
  b. weak forces.
  c. adhesive forces.
  d. gravitational forces.

Part 3. What Would Happen If? Explaining What Happened

Directions: Explain in words or with drawings the behavior of water in the following situations. Use scientific vocabulary appropriately. Some of the vocabulary terms you may need are listed below.

Vocabulary: surface tension, adhesion, adhesive forces, cohesion, cohesive forces, atomic particles, and attractive forces

Situation 1: A small cup of water is filled to almost, but not quite, overflowing. A paper clip is carefully dropped into the cup. The water does not overflow until fifty or more paper clips are added to the cup. **Explain why the water doesn't overflow as paper clips are added.**

Situation 2: A student carefully places drops of water on two different surfaces: 1) wax paper and 2) plastic wrap. The water drops bead on wax paper and spread out on plastic wrap. **Explain why water beads on one surface and spreads out on a different surface.**

Situation 3: Pepper is sprinkled on the surface of a small bowl of water. Next, a drop of liquid soap is added to the water. As the liquid soap is added to the water, the pepper is quickly pushed to the side of the container. **Tell why liquid soap spreads the pepper.**

Teacher Notes: Post-assessment – Attractive Forces
Lesson Thirteen: Teacher Scoring Guide

Part 1. Agree or Disagree

1. (A) As more and more paper clips are added to an already full cup of water, the water dome on the top of the cup gets larger and larger.

2. (D) **Cohesion describes the attraction between unlike molecules. For example, cohesive forces cause water to spread out on plastic wrap.** Adhesion is when unlike molecules are attracted to each other. For example, cohesive forces cause water to spread out on plastic wrap. Or, cohesion is when like molecules are attracted to each other. For example, cohesive forces cause water to form drops.

3. (A) Adding paper clips to a full cup of water created an upward force that eventually broke the water's surface tension, and water overflowed the cup.

4. (D) **Liquid dish detergent increased the surface tension of water.** Liquid dish detergent **decreased** the surface tension of water. Fewer paper clips were added before the surface tension was broken.

5. (A) Temperature can affect surface tension.

6. (D) **Surface tension is a characteristic of solids, liquids, and gases.** Surface tension is a characteristic property of **liquids only.**

7. (D) Water molecules strongly adhere to wax molecules. Water molecules **weakly adhere to wax molecules.** Water beads on wax paper. Water molecules seem to repel wax particles.

8. (A) Detergent molecules adhere to both water and fat molecules.

9. (D) **Forces on molecules at the surface of a liquid are balanced as compared to forces on water molecules in the middle of a liquid.** Forces on molecules at the surface of a liquid are **unbalanced.**

10. (A) Attractive forces between molecules can explain the behavior of matter.

Part 2. Multiple Choice.

Best Responses: 1=B, 2=B, 3=A, 4= D, 5= A, 6= C

1. Surface tension is a characteristic property of
    a.   only solids.
    b.   only liquids.
    c.   only gases.
    d.   solids, liquids, and gases.
2. Cohesion is best described as the
    a.   attraction between pepper and water.
    b.   force that holds molecules of the same substance together.
    c.   sticking together of two different substances.
    d.   way that liquid molecules move about in liquids.
3. A surfactant, like dishwashing liquid,
    a.   reduces surface tension.
    b.   increases surface tension.
    c.   preserves substances.
    d.   cools substances.
4. Which is the best explanation for why oil and water do not mix?
    a.   Oil particles are slippery.
    b.   Water particles are only attracted to other water particles.
    c.   Oil is denser than water.
    d.   Oil particles are not attracted to water particles.
5. When a liquid is cooled, the surface tension
    a.   increases.
    b.   decreases.
    c.   remains the same.
    d.   cannot be measured.
6. Attractive forces between water molecules and plastic wrap are best described as
    a.   cohesive forces.
    b.   weak forces.
    c.   adhesive forces.
    d.   gravitational forces

Evaluation Suggestion
    Advanced: 14 -16
    Proficient: 11-13
    Partially Proficient: 8-10
    Unsatisfactory: 0-7

Part 3. What Would Happen If? Explaining What Happened
Vocabulary: surface tension, adhesion/adhesive forces, cohesion/cohesive forces, surfactant

Scoring Criteria
Advanced: Explains all three situations accurately; includes key points; uses scientific vocabulary appropriately in explanations.
Proficient: Explains two or three situations adequately; includes key points; may not use vocabulary but adequately describes or draws concept.
Partially Proficient: Explains one of the three situations adequately; may not use vocabulary but adequately describes or draws the concept.
Unsatisfactory: Unable to explain any of the situations. Shows little understanding of concepts.

Situation 1: A small cup of water is filled to almost, but not quite, overflowing. A paper clip is carefully dropped into the cup. The water does not overflow until fifty or more paper clips are added to the cup. **Explain why the water doesn't overflow as paper clips are added.**

*Water has a **strong surface tension**. The **cohesive force** between individual molecules of water is very strong. As paper clips are added, the water's surface tension holds the water molecules in the cup until too much pressure is placed on the surface. This breaks the attractive forces between the water's surface molecules and the water overflows the cup.*

Situation 2: A student carefully places drops of water on two different surfaces: 1) wax paper and 2) plastic wrap. The water drops bead on wax paper and spread out on plastic wrap. **Explain why water beads on one surface and spreads out on a different surface.**

*Water molecules are not strongly attracted to wax molecules but they are attracted to plastic wrap molecules. The **adhesion** between water and wax is less than the adhesion between water and plastic. Thus, water beads on wax and spreads out on plastic.*

Situation 3: Pepper is sprinkled on the surface of a small bowl of water. Next, a drop of liquid soap is added to the water. As the liquid soap is added to the water, the pepper is quickly pushed to the side of the container. **Tell why liquid soap spreads the pepper.**

*Pepper floats on water's surface. It is does not break the surface tension. Soap is a **surfactant**. It is able to break the attractive forces between water molecules and spread out on the surface. When soap is added, the pepper is pushed to the sides of the container as the **soap breaks the water's surface tension**. Pepper is not strongly attracted to soap so it stays to the edges or falls to the bottom of the container.*

Literacy Resource Section

The literacy strategies used in *Synergized Middle School Chemistry (SMSC)* have a research base and are classroom tested. Following the chart is an alphabetical listing of the strategies with brief descriptions and references for further instructional information.

Literacy Strategies Focus: Vocabulary & Reading Comprehension	Lessons MP = Matter's Phases D = Density AF = Attractive Forces	Lesson Titles
•Shared Reading – Everybody Read To…(ERT)	MP: Lesson Two	Reading: Chemical Safety and Safety Rules
•Student Investigation-Before Reading •Word Splash	MP: Lesson Four	Reading: A Solid Start
•Student Investigation-Before Reading •Question-Answer Relationships (QAR) Shared Reading •Student Investigation-After Reading	MP: Lesson Seven	Reading: Phasing Into Liquids
•Visual Representation (Student-made Poster)	MP: Lesson Eleven	Air Matters
•Student Investigation-Before Reading •Semantic Feature Analysis	MP: Lesson Thirteen	Reading: Comparing Matter's Phases
•Student Investigation-Before Reading •Most Important Words •Three-Part Vocabulary	MP: Lesson Fifteen	Reading: Diffusion
•Student Investigation-Before Reading •Anticipation Guide	MP: Lesson Seventeen	Reading: Boiling Hot or Boiling Cold?
•Chart Reading and Data Interpretation	MP: Lesson Nineteen	Graphing Phase Changes
•Word Sort with Concept Map •First Word – Last Word	MP: Lesson Twenty	Reviewing Matter's Phases
•Student Investigation-Before Reading •Question-Answer Relationships (QAR) •Semantic Map	D: Lesson Three	Reading: Defining Density
•Student Investigation-Before Reading	D: Lesson Four, Part 2	Reading: What About the Density of the Grape
•Chart Reading and Data Interpretation	D: Lesson Six	Making Sense of Density Data
•Semantic Map	D: Lesson Nine	Revisiting the Density Semantic Map
•Student Investigation-Before Reading •Question-Answer Relationships (QAR) Shared Reading	AF: Lesson Three	Reading: Water's Amazing Invisible Elastic Skin
•Concept Definition Map	AF: Lesson Five	Surface Tension
•Student Investigation-Before Reading •Main Idea Map	AF: Lesson Seven	Reading: May the Forces Be With You
•Student Investigation-Before Reading •Anticipation Guide	AF: Lesson Ten	Reading: From Milky Swirls to Washing Clothes
•Chart Reading and Data Interpretation	AF: Lesson Eleven	Surface Tension Measurements

Synergized Middle School Chemistry

Anticipation Guide

> Anticipation guides (Herber, 1978) are a set of carefully selected questions that serve as a pre/post inventory for a reading selection (Barton & Jordan, 2001, p. 72).

This easy-to-use strategy motivates students to read nonfiction text for specific information. Adolescents struggle with nonfiction text because they lack the background and reading skills to know "what's most important" in the text. To guide their reading, the teacher develops five to ten statements (depends on length of text) to give to students before they read. Some statements agree with the text and other statements are in disagreement with the text. Before reading, students work in pairs to determine whether they agree or disagree with each statement. As students discuss the statements, they are introduced to key vocabulary. After a discussion of each statement, students read to determine what the text says about each statement. After reading, students reread the statements and make corrections based on the text. Because anticipation guides are self-correcting, there is a high probability that all students will focus on the key information in the text and feel successful about their reading.

Concept Definition Mapping

> Concept definition mapping (Schwartz, 1988) is a strategy for teaching students the meaning of key concepts (Barton, 2001, p. 50).

This strategy builds conceptual understanding by using a graphic organizer to relate a key concept to a bigger idea, concept examples, and the concept's key features. Students are prompted to read the text and complete the concept map during reading. After reading, students pair-share concept maps and any additions or corrections are made. Once maps are completed, students can use their maps to write a definition for the concept in their own words. (See completed graphic organizer in Teacher Notes, Attractive Forces, Lesson Five.) Concept definition maps are reserved for *key* concepts. The maps can be kept for reference and/or displayed in the classroom.

First Word – Last Word

> Research by Stahl and Fairbanks (1986) indicates that students' achievement may increase as much as 33 percentile points when vocabulary instruction focuses on specific words that are important to what students are learning (Barton & Jordan, 2001, p. 18).

This challenging, yet fun, crossword game focuses students on *key* words and concepts. It is an excellent way to review terms at the end of an instructional sequence. For this strategy, the teacher selects a key word or phrase and writes it vertically for the class to see. (See Matter's Phases: Lesson Twenty for an example.) Students then use their notes and other resources to find words from the previous lessons that start with each letter. For example, if the teacher selected the word "matter", students would be searching for related words that start with "m", "a", "t", "t", "e", and "r". They might select "mass" for "m". When students are finished, they share their words and connections with the class and write sentences relating each of the words to the *key* word.

Main Idea Map

> ...nothing is sacred or irreplaccable about the actual formula and format for traditional outlining. What is irreplaceable is the essential need for organizing information (Hyerle, 1996, p. 53).

A main idea map is a task specific graphic organizer used during reading. (See completed idea map in Teacher Notes, Attractive Forces: Lesson Seven.) As students read, they first determine "the most important" idea or main idea of the reading as a whole. Next, they read to find out the "important ideas" that support the main idea. Finally, they add "details" to the "important ideas." When they are finished, they have a hierarchical organizer (outline) of the text in a student friendly format to use for future lessons and reference.

The Most Important Words

> ...science reading is an interactive-constructive process by which they (students) make meaning from personal experience, recorded experiences of other people, and the context of reading (Barton, 2001, p. 30)...

This vocabulary and comprehension strategy places the responsibility for selecting *key* words in the hands of students. Before reading, the teacher prompts the students to select a specific number of words (based on length and text subject) to declare the most important words in the selection. After reading, they pair-share their words. Each pair then shares their words with the class, and the class comes to agreement on the most important words.

Visual Representation (Student-made Poster)

> By using visual tools, a teacher simultaneously has access to and may assess both the "content" knowledge and the "processes" that a student is employing to come to a new understanding. Is the student's content knowledge weak, or is the thinking and reasoning problematic? Or both? (Hyerle, 1996, p. 126)

The goals of student-created visual representations are for students to think more deeply about concepts and events, build on their previous knowledge and experiences, develop new understandings, and to apply their literacy skills as they create their final product. Before students begin their work, the teacher states the purpose of the visual, the elements to include, and the way the visual will be used (e.g., display only, oral presentation, and/or performance assessment). If it is to be used as an assessment, scoring criteria or a rubric should be provided.

Question-Answer Relationships (QAR)

This strategy (QAR) focuses on the relationship between questions and answers. It teaches students that addressing different kinds of questions requires different thought processes (Barton & Jordan, 2001, p. 117).

Students are presented with four different levels of questions about a reading:
1. *Right There* questions have their answers found in the same sentence of the text,
2. *Think and Search* questions relate ideas in a passage or paragraph and involve inferring skills,
3. *Author and You* questions combine a students prior knowledge with the text, and
4. *On My Own* questions are answered from the student's background and are not found in the text. (Raphael, 1982, 1984, 1986)

These questions can be presented as prompts during a shared reading or given to students to answer as they read on their own or collaboratively. These types of questions help students understand that reading involves interacting with text and not just reading the words. They learn that authors (and teachers) expect them to use their own prior knowledge and experience to make text more meaningful.

Semantic Feature Analysis

Semantic feature analysis (Baldwin, Ford, & Readance, 1981; Johnson & Pearson, 1984) helps students discern a term's meaning by comparing its features to those of other terms that fall into the same category (Barton & Jordan, 2001, p. 58).

Feature analysis charts are an excellent way to compare and contrast similar concepts. Using a matrix, students organize text information by relating key concepts to their features. Concepts are listed as row headings and features are listed as column headings. For example, in Matter's Phases: Lesson Thirteen, the phases of matter are listed as row headings and the molecular behavior of atoms and molecules in each phase are listed as features.

Semantic Map

A semantic map is a visual tool that helps readers activate and draw on prior knowledge, recognize important components of different concepts, and see the relationships among these components (Barton & Jordan, 2001, p. 61).

After students have studied a concept in depth, a *key* word (or phrase) is written on the board. Students then brainstorm all the terms they know that connect to the concept. On a sheet of chart paper, students write and circle the key word and then group the brainstormed words/phrases into categories. These categories are mapped or webbed to the key word. Maps/webs are displayed and discussed. Semantic maps can be started early in a lesson set to introduce a key concept and then revisited at a later date to evaluate new conceptual understandings.

Shared Reading – Everybody Read To... (ERT)

Round robin reading is not a multilevel activity in which everyone feels successful, nor can multiple things be learned! ERT..., that is, Everybody Read To..., is a way of guiding the whole class (or small group) through reading a selection. We use ERT (which rhymes with *hurt*) when we want students to do an initial reading on their own but also want to keep them together to provide guidance and support (Allington, 2007, p. 186).

Shared reading, a form of guided reading, is an excellent way to teach nonfiction comprehension skills to a heterogeneous group of students. Using a multilevel text, the teacher and students read together. The Everybody Read To ... or ERT strategy works well for middle level students. The teacher begins by prompting students to read a portion of the text "to find out" or "to figure out" the answer to a question. Questions focus on text features, key words, and concepts. When students think they have found the answer in the reading, they raise their hands. The teacher waits until most hands are raised before asking a student with a raised hand for a response. In ERT, one student gives a response and a second student is called on to tell where the information is found in the reading. Allington reports that ERT allows students to be successful in a multilevel lesson with multiple learning goals.

Student Investigations

For students to make sense of what they read, they need to be able to grasp and make sense of new information in light of what they already know. When readers activate and use their prior knowledge, they make the necessary connections between what they know and new information. Teachers should help students recognize the important role that prior knowledge plays and teach them to use that knowledge when learning science through reading (Barton and Jordan, 2001, p. 3).

All reading lessons should have a "before reading" strategy that prepares students for the reading by activating prior knowledge. In science, hands-on investigations "level the playing field" for all students by providing common background experiences to help make sense of the reading content. Investigations also motivate students to read about a topic by providing a purpose for reading. Vocabulary should be introduced in the context of the investigation; however, formal definitions can wait until the student has read about the concepts. In some situations, another investigation or demonstration may follow the reading as an "after reading" strategy. These "after reading" investigations apply and reinforce concepts presented in the readings.

Three-Part Vocabulary

Although science teachers may be aware of the need for students to learn science terms and phrases, they may not know the most effective ways to help students learn these words. Consequently, many teachers resort to teaching vocabulary the way they were taught: looking words up in a dictionary and memorizing their definition (Barton & Jordan, 2001, p. 14).

Synergized Middle School Chemistry

Three-part vocabulary is a strategy that helps students make their own meaning of scientific terms by exposing students to the terms in a variety of different ways. Terms are introduced in the context of an investigation. The teacher, aware of the key vocabulary, continues to use the terms during informal and structured class discussions. Next, students read an article that uses the terms. During and after reading strategies focus on the term's use in the context of the reading. After students have had multiple exposures to the terms, they compile a three-part vocabulary chart. This chart is then used as a reference for later lessons.

Students divide their paper into three columns. In the left-hand column they write the key word or phrase, in the middle column they write a definition *in their own words*, and in the right-hand column they draw a picture illustrating the word. Their illustration can come from their previous experiences inside or outside the classroom. Word charts can be added to a vocabulary section in their science notebook or placed on index cards for future reference. Only key words and phrases should be used for three-part vocabulary.

The science notebook is an excellent place for students to record their key vocabulary with definitions and examples. One method is to designate the end of the notebook for a personal science dictionary. Start with the last page and work forward.

Word Sort

> A word sort is a simple yet valuable activity…The object of word sorting is to group words into different categories by looking for shared features among their meanings (Vacca & Vacca, 2002, p. 175).

This strategy is used to build meaningful connections between technical terms at the end of an instructional sequence. In a word sort, the teacher selects the words. Words are written on cards. In a closed-sort, students are given categories and asked to sort the words into the categories and provide reasons for their choices. In an open-sort, students are given words, and they create the categories. After sorting the words, students share how they grouped their words. After the sort, students summarize what they have learned. This can be done in a concept or semantic map or by writing summary sentences. A word sort is an excellent pre-writing strategy.

Word Splash

> *Teachers can help students build conceptual knowledge of content area terms by teaching and reinforcing the concept words in relation to other concept words.* This key instructional principle plays itself out in content area classrooms whenever students are actively making connections among the key words in a lesson or unit of study (Vacca & Vacca, 2001, p. 165).

This is a favorite vocabulary building strategy of both teachers and students. Using the strategy, a difficult or not so interesting reading becomes much more exciting as students search for word connections. First, the teacher selects ten to twenty words (depending on the type and length of reading) to "splash" on a page. (See Matter's Phases, Lesson Four, for an example of a splash created with Microsoft Word Art.) Before reading, students work in pairs to draw connection lines between the words. On each line, students write a word or phrase that describes the connection. After all pairs have finished, every word is discussed and possible connections explained.

Next, students read the article. During reading, they search for the words and how they are connected. Usually, students start by scanning the article for the words. This is a good strategy; however, they usually find out that they must read and reread the article a number of times to make all the connections. They learn that reading nonfiction text usually requires more than one read through.

After reading, the reading pairs compare and contrast word connections to their pre-reading predictions. Students pair-share their new connections with the class. Each student then writes sentences that connect all the words.

Writing Strategies

Science notebook entries provide opportunities for writing in various modes while learning science. While these nonfiction modes should be explicitly taught during language arts, they are reinforced in the context of doing science, giving meaning and purpose to the writing (Kotelman, Saccani, and Gilbert, 2006, p. 151).

In their 2006 article *Linking Science and Literacy in the K-8 Classroom*, Kotelman, Saccani, and Gilbert describe five modes of writing reinforced and practiced in science classes: descriptive, explanatory, procedural, recount, and persuasive. *SMSC* lessons provide many opportunities for science teachers to reinforce these writing modes.

SMSC investigation prompts (e.g., predictions, observations, claims and evidence and new understandings) require students to describe their work, explain their thinking, detail their procedures, and use evidence to support their claims.

In addition, many of the literacy strategies are excellent writing prompts. For example, concept and semantic maps, word sorts, and word splashes can be used for descriptive and explanatory writing. Summarizing or recounting is a valuable science literacy skill. *SMSC* lessons frequently ask students to summarize their investigation findings and their readings.

Literacy Strategy References

Allington, Richard L., and Patricia M. Cunningham. 2007. *Classrooms that work: they can all read and write.* New York, NY: Pearson.

Barton, Mary Lee, and Jordan, Deborah. 2001. *Teaching reading in science.* Aurora, CO: McREL. Available online from the Association for Supervision and Curriculum Development at ASCD.org Store.

Baldwin, R. S., J. C. Ford, and J. E. Readance. 1981. Teaching word connotations: An alternative strategy. *Reading World* (21): 103-108.

Herber, H. 1978. *Teaching reading in content areas* (2nd ed.). Englewood Cliffs, NJ: Prentice Hall.

Johnson, D. D., and P. D. Pearson. 1984. *Teaching reading vocabulary* (2nd ed.). New York: NY: Holt Rinehart and Winston.

Kotelman, Marleen, Toni Saccani, and Joan Gilbert. 2006. Writing to learn: science notebooks, a valuable tool to support nonfiction modes/genres of writing. In *Linking science & literacy in the K-8 classroom*, eds. R. Douglas, M. P. Klentschy, and K. Worth with W. Binder, 149-161. Arlington, VA: NSTA Press.

Raphael, T. E. 1982. Question-answering strategies for children. *The Reading Teacher* (36): 186-190).

Raphael, T. E. 1984. Teaching learners about sources of information for answering comprehension questions. *Journal of Reading* (27): 303-311.

Raphael, T. E. 1986. Teaching question-answer relationships, revisited. *The Reading Teacher* (39): 516-522.

Schwartz, R. 1988. Learning to learn vocabulary in content area textbooks. *Journal of Reading* (32): 108-117.

Stahl, S. A., and M. M. Fairbanks. 1986. The effects of vocabulary instruction: A model-based meta-analysis. *Review of Education Research* 56 (1): 72-110.

Vacca, Robert T., and Jo Anne L. Vacca. 2002. *Content area reading: Literacy and learning across the curriculum* (7th ed.). Boston, MA: Allyn and Bacon.

References

Doran, Rodney. F. Chan, P. Tamir, and C. Lenhardt. 2002. *Science educator's guide to laboratory assessment*. Arlington, VA: NSTA Press.

Educational Innovations, Inc. 362 Main Avenue, Norwalk, CT 06851 www.teachersource.com

Flinn Scientific Inc., P.O. Box 219, Batavia, Illinois 60510-0219 www.flinnsci.com

Marzano, Robert J., D. Pickering, and J. Polluck. 2000. *Classroom instruction that works*. Alexandria, VA: Association for Supervision and Curriculum Development.

National Research Council (NRC). 1996. *National science education standards*. Washington: DC: National Academy Press.

National Research Council (NRC). 2001. *Classroom assessment and the national science education standards*. Washington: DC: National Academy Press.

National Research Council (NRC). 2001. *Inquiry and the national science education standards*. Washington: DC: National Academy Press.

Institute for chemistry Literacy through Computational Science: Chemistry Simulations. 2010. Retrieved November 9, 2010, from http://iclcs.illinois.edu/index.php/chemistry-simulations.

American Chemical Society, Science Safety Guidelines. 2010. Retrieved November 9, 2010, from http://portal.acs.org

Made in the USA
Monee, IL
28 August 2023

41780240R00122